I0476627

NCA Report Series, Volume 7

Climate Change Modeling and Downscaling:

Issues and Methodolgical Perspectives for the U.S. National Climate Assessment

NCA Report Series

The National Climate Assessment (NCA) Report Series summarizes regional, sectoral, and process-related workshops and discussions being held as part of the third NCA process.

This workshop focused on questions, issues, and methodological perspectives regarding the use of mathematical models for the NCA, as well as the complexities that arise when applying data and models to multiple spatial and temporal scales. The workshop was held in Arlington, VA on December 8-10, 2010. Volume 7 of the NCA Report Series summarizes the discussions and outcomes of this workshop. A list of planned and completed reports in the NCA Report Series can be found online at http://assessment.globalchange.gov.

CONTENTS

CONTENTS

PREFACE

In December 2010, the Office of Biological and Environmental Research of the U.S. Department of Energy in conjunction with the U.S. Global Change Research Program convened a workshop in Arlington, Virginia to address questions and issues regarding the use of mathematical models for the National Climate Assessment (NCA). The workshop also addressed the complex issues that arise when data and models are applied at multiple spatial and temporal scales, an inherent requirement of a national climate assessment. An Interagency National Climate Assessment (INCA) Task Force provided guidance regarding purpose, scope, and objectives of the workshop. The results of the workshop in turn offer guidance not only for the National Climate Assessment but also more generally for future research in the use of models for these tasks.

The workshop included plenary-session presentations and panels as well as breakout group discussions. Presentations and discussions covered both global and regional climate models and linkages of these models to enable regional analyses or downscaling of variables constrained by global model results. Participants reviewed potential use of carbon cycle and other biogeochemical cycling models as well as a range of models for analyzing the impacts of climate change on various sectors and within different regions. Integrated assessment models that relate climate change to human factors were discussed, as were various options for coupling models (used to assess climate change impacts) to either data on human activities or models of various human systems. Participants offered significant insights about the use of models within the overall NCA structure of linked sectors and regions. A number of workshop findings and recommendations apply to the development of a longer-term, sustained Assessment process.

Prior to the workshop, a small team (with experience in the use of models for climate change assessments and the associated issues imposed by the need to address multiple temporal and spatial scales) prepared a white paper to inform assessment planning and stimulate discussion within the assessment community. Workshop planners and participants found the white paper useful for background information and for organizing discussions during the meeting. Following the meeting, workshop organizers prepared a letter report summarizing key findings and recommendations.

This full report of the workshop includes three chapters. Chapter 1 is the letter report and can be read as an executive summary of the full document. Chapter 2 is the white paper, and Chapter 3 reports the workshop proceedings, findings, and recommendations. Collectively, the chapters provide a comprehensive summary of not only the workshop proceedings but also the background material considered by meeting participants. This report does not reflect a consensus by participants, but rather describes requirements, options, and challenges as perceived by individual participants in the workshop setting and in their contributions to the background white paper.

Appendices include the workshop agenda and a participant list.

Chapter 1:
Executive Summary - Modeling and Scaling Issues in the National Climate Assessment

Anthony C. Janetos, Pacific Northwest National Laboratory; William Collins, Lawrence Berkeley National Laboratory; Don Wuebbles, University of Illinois; Noah Diffenbaugh, Stanford University; Katharine Hayhoe, Texas Tech University; Kathy Hibbard, Pacific Northwest National Laboratory; George Hurtt, University of Maryland

1.1 Introduction

This chapter provides a brief background about the U.S. National Climate Assessment (NCA) and describes the primary findings and recommendations of the workshop on Climate Change Modeling and Downscaling that occurred December 8–10, 2010 in Arlington, Virginia. The U.S. Department of Energy sponsored the meeting. The material in this chapter of the workshop report was assembled initially as a letter report of the meeting to the National Climate Assessment Development and Advisory Committee (NCADAC – a federal advisory committee), and as such provides a brief summary of the workshop background, proceedings, and findings. Two subsequent chapters address these topics in more detail.

1.2 Background

The National Climate Assessment (NCA) 2013 report is required under the Global Change Research Act of 1990, which says in Section 106 that a "Scientific Assessment" must be prepared not less frequently than every four years and delivered to the President and Congress. This Assessment must

- Integrate, evaluate, and interpret the findings of the U.S. Global Change Research Program (USGCRP), and discuss the scientific uncertainties associated with such findings;
- Analyze the effects of global change on the natural environment, agriculture, energy production and use, land and water resources, transportation, human health and welfare, human social systems, and biological diversity; and
- Analyze current trends in global change, both human-induced and natural, and project major trends for the subsequent 25 to 100 years.

This last requirement to analyze trends into the future requires the use of models at various scales and also the ability to build scenarios that help describe and analyze future conditions where changes in climate are only one of a myriad of changing variables.

Assessments serve important functions by providing the scientific underpinnings of informed policy. They also serve as progress reports by identifying advances in the underlying science, providing critical analysis of issues, highlighting key findings and key unknowns that can improve policy choices, and guiding decision making related to climate change.

The approach that is envisioned for the third NCA report is a comprehensive assessment of climate change, impacts, vulnerabilities, and response strategies within the context of how communities and the Nation as a whole work to create sustainable and environmentally sound development paths.

A primary goal of the new NCA is to establish a permanent assessment capacity both inside and outside of the federal government. This new NCA will be an ongoing process that draws upon the work of stakeholders and scientists across the country. Assessment activities will result in the capacity to do ongoing assessments of vulnerability to climate stressors, observe and project impacts of climate change within regions and sectors, develop consistent indicators of progress toward reducing vulnerability, and allow for the production of a set of reports and Web-based products that are useful for decision making at multiple levels.

Strategic planning for the new NCA began in early 2010. A draft strategic plan has been developed within an Interagency National Climate Assessment (INCA) Task Force that includes members from all 13 USGCRP agencies plus several others. The NCA will continually solicit input from a broad range of stakeholders, decision makers, and concerned citizens to ensure that its vision and implementation are responsive to their needs. The NCADAC will review research results and technical inputs prepared by interested organizations, groups, and individuals; oversee the Assessment process; and prepare the 2013 Assessment report.

1.3 Goals and Objectives of the NCA

The vision for the National Climate Assessment is to establish a continuing, inclusive national process that

- Synthesizes relevant science and information;
- Increases understanding of what is known and not known;
- Identifies needs for information related to preparing for climate variability and change and reducing climate impacts and vulnerability;
- Evaluates progress of adaptation and mitigation activities;
- Informs science priorities;
- Builds assessment capacity in regions and sectors; and
- Builds societal understanding and skilled use of assessment findings.

Although the definition of regions to be used in NCA 2013 was under consideration at the time of the workshop, it was noted that the ability to deploy information on the Web will significantly relieve the pressure to define regional boundaries. If the NCA can provide information at a number of national, regional, and local scales, the exact boundaries of the regions become less important. However, at a recent regional and sectoral workshop there appeared to be consensus for regions roughly analogous to those defined for the 2009 report, with adjustments to use state boundaries wherever possible.

There is a strong desire for both understanding regional climatology and having the capacity to project conditions at the regional level at multiple temporal scales, including seasonal to inter-annual, decadal, and 50–100 years. The need to understand change in both a transient and end-point framework was also noted.

1.4 The Role of Modeling in the NCA

The charge for the NCA, as laid out in the Global Change Research Act, requires an assessment of potential impacts and trends 25 to 100 years into the future. This charge raises significant issues for the modeling communities that will be involved in the NCA. It requires that both physical and economic models not only be used in a way that captures our current understanding of climate change, its impacts, and possible response strategies, but also to provide forward projections. Thus, there is a clear need for the use of climate models, impacts models, and integrated assessment models. The NCA will need to be completely transparent about the abilities of various models used to do projections; where there are measurement, parameterization, or structural uncertainties; and the degree to which those uncertainties affect the confidence in findings. There is also the fundamental need for the development and application of fully-coupled models of human and natural systems.

1.4.1 Climate Models
A range of tools is used to study potential changes in the climate system, including global climate models, regional-scale climate models, and Earth system models of intermediate complexity. High-resolution climate models allow for improved representation of fine-scale climate processes, including those associated with topography, land-cover change, and the carbon cycle, as well as mixing and circulation in the atmosphere and ocean. Because these processes can regulate the response of local and regional climate to elevated greenhouse gas concentrations, it is critical that assessments of the potential impacts of climate change consider fine-scale phenomena.

Regional treatments in global climate models, whether nested or offline calculations, offer an opportunity to conduct targeted high-resolution experiments. However, nested experiments are still computationally demanding, particularly given the many dimensions of climate change uncertainty space that must be probed. For instance, high resolution is necessary to capture climate processes at local scales that ultimately inform climate change impacts. Therefore, an extant climate domain is necessary to capture the influence of fine-scale processes and thereby project local-scale changes across a large geographic area. Alternatively, ensemble methods that generate multiple model realizations of a single experiment allow for quantification of the likelihood that simulated climate changes are due to prescribed changes in climate forcing (e.g., changes in atmospheric greenhouse gas concentrations) rather than to internal model variability.

In the majority of impact studies, results derived using global climate models are typically downscaled from coarse spatial resolution (e.g., 1–3° latitude × longitude) to a resolution that is relevant to the process or mechanism that the impact model is addressing, generally at local or regional scales (e.g., 0.1–0.3° latitude × longitude). This can be done either using a regional climate model (with boundary conditions from a global climate model) or by using statistical downscaling techniques that combine modeling results from global or regional climate models with observational records. For decision making or for regional evaluation of the consequences of changing climate, the nature of downscaled global climate model information to initialize or drive impact models imposes limitations. We need to understand the variability in downscaled information that derives from differences in (1) global climate models from which the information is derived; (2) the methods by which the downscaled results were obtained; and (3) how the downscaled results sample the entire range of relevant climate variability, including the tails of the distributions.

1.4.2 Impact Models

Impact models have been used to evaluate and assess the consequences of changing climate to human and environmental systems. Impact models comprise multiple aspects of biogeochemical, biophysical, rule-based, conceptual, and integrated responses to climate through various techniques including process and mechanistic simulation, statistical and empirical relationships, and time series analyses. There are many different types of models that have been used to assess climate impacts: ecosystem models, which calculate fluxes of materials and energy within and to and from ecosystems and are generally used to evaluate the ways in which such system variables such as net primary productivity respond to changes in climate; hydrologic models, which simulate water flow and quantity at a variety of spatial scales; dynamic vegetation models, which simulate disturbance and long-term trends in vegetation geographic distributions as a function of change in the climate system; and agricultural productivity models, which simulate changes in agricultural production. These are just a few of the many types of models that have been used in the past in climate impact assessments, and they will also have a role to play in the current National Climate Assessment.

Since impacts that result from climate and or land use change can in turn affect future climate, there is increased recognition of the need to couple predictions of impacts back into climate calculations. Previous U.S. assessments (e.g., NAST, 2001) and the U.S. Climate Change Science Program Synthesis and Assessment Products provide information about how climate change could impact productivity, quality, and vulnerability within various sectors (e.g., forests, rangeland, water, livestock production, ecosystem, coastal, marine, human health, and transportation systems), with an acknowledgement of these limitations.

1.4.3 Integrated Assessment Models

Integrated assessment models simulate the interactions between human decisions in the energy sector and both carbon and physical climate systems as components of an overall Earth system approach. Historically, they have been used to simulate greenhouse gas emissions trajectories as a function of different mitigation targets, usually over the 21st century. They include the major interactions with other components of the Earth system, and most include a reduced form model that can be used to calculate global radiative forcing, if desired. All such models have representations of agricultural productivity to meet food demands. There are only a small number of sophisticated integrated assessment models in existence, although more are being developed.

There are differences in the degree of foresight that integrated assessment models assume in their simulations, in the delineation of regions for analysis, and in the degree of specificity with which they describe energy technologies. A critically important feature to note about integrated assessment models is that although they are based on observations of energy and agricultural variables as well as relationships between energy production and greenhouse gas emissions, they are not used to predict societal decision making. Rather, they are used to evaluate the potential consequences of different decision trajectories, usually with respect to mitigation targets.

As integrated assessment models have developed, many now also include land-use decisions (e.g., forestry) as part of their analysis—at least that component of land use that is directly related to meeting demands for food and that interacts in various ways with the energy system (Wise et al., 2009). These relationships are captured in the model results being used in the fifth assessment report of the Intergovernmental Panel on Climate Change (IPCC); for example, in their calculations of the emissions and greenhouse gas concentration trajectories - known as representative concentration pathways (RCP's) (Moss et al., 2010; Thomson et al., 2010).

1.5 Issues of Importance for the NCA

There are many ways in which the various families of models that will play a role in the NCA will be improved over the next several years and decades. A critically important feature of the workshop was to focus on those issues that are of the highest importance to the NCA itself. We identify six such issues below, and summarize the workshop participant comments about these:

1. How will the output from global climate models be used to describe local and regional climate projections in the NCA?
2. How will emissions scenarios be defined and used in the NCA? How are these scenarios related to scenarios of vulnerability in both human and natural systems, and how will those be modeled?
3. What is the relationship between Phase 5 of the Coupled Model Intercomparison Project

(CMIP5) efforts at global model intercomparisons and model output that will be used in the NCA?

4. How will inter-sectoral impacts be modeled in the NCA, and which inter-sectoral impacts are key to informing regional decision making in the context of climate?

5. How will model uncertainties be characterized and quantified in the NCA?

6. How might integrated assessment models be used in the NCA?

For each of these questions, the following sections summarize the discussions and workshop recommendations. These are not consensus opinions of all workshop participants but represent the best judgment of the authors (derived from workshop deliberations) as to options that the NCA and its federal advisory committee might consider.

How will the output from global climate models be used to describe regional climate projections in the NCA?

The workshop and white paper outline several possible methods for downscaling global model output for use in regional studies and models. As a result of those discussions, there are several options for the NCA to consider:

- Use existing global model runs from Program for Climate Model Diagnosis and Intercomparison (PCMDI), focusing most on the Phase 3 Coupled Model Intercomparison Project (CMIP3) archive, and downscale the results using one of many possible regional models or through subsequent statistical analyses. The specifics of which model to use and which variables to focus on will depend on the representativeness of the modeling tools and the particular analyses desired for each region.

- Use existing regional model runs from the North American Regional Climate Change Assessment Program (NARCCAP) and other research programs to form the basis for subsequent analyses. These have been done using global runs from CMIP3 to provide initial conditions from a selection of global climate models. Thus, the approach could be viewed as a variant of the first option, but not necessarily requiring new climate modeling to be done. However, these model results need to be carefully scrutinized as part of the analysis process.

- Construct a unified national, high-resolution data set with either statistical downscaling or dynamic regional models, but do not attempt

to tailor it either to specific impact studies or to specific regions. This is basically the procedure that was used in the first national assessment.

How will emissions scenarios be defined and used in the NCA? How are these scenarios related to scenarios of vulnerability in both human and natural systems, and how will those be modeled?

There was extensive discussion of emissions scenarios in the context of questions about using the CMIP3 family of models versus the CMIP5 family of global models. Part of the reason for this discussion was that the scenarios described by the IPCC Special Report on Emissions Scenarios (SRES) (Nakicenovic and Swart, 2000), while providing a range of possible emissions futures, do not include explicit stabilization pathways. The CMIP5 scenarios that use the IPCC Representative Concentration Pathways (RCPs) do include explicit stabilization scenarios at 4.5 watts per square meter (W/m^2) and 6 W/m^2 as well as an overshoot and stabilization at 2.6 W/m^2. There was concern voiced that because both Working Groups I and II of the IPCC fifth assessment process plan to report on results based on CMIP5, that the NCA ought to have some way to provide comparability in terms of atmospheric concentrations and greenhouse gas emissions, since the initial reporting from the NCA will be so close in time to the planned release of the IPCC reports. This is explored in greater detail below.

What is the relationship between the CMIP5 efforts at global model intercomparisons and model output that will be used in the NCA?

A majority of workshop participants felt that while it would be desirable to use model results from CMIP5 in order to maximize the ability to do comparisons with the IPCC fifth assessment, the simulations themselves are very unlikely to be available in time for full use in the NCA. In addition, the global models that are included in CMIP5 intercomparisons are new versions of existing atmosphere-ocean general circulation models, with some model runs at higher resolution than for those involved in the fourth IPCC assessment, and many of the new models have incorporated new atmospheric physics and an active carbon cycle, and in some cases the nitrogen cycle as well. The performance of these new models will be less well understood for some time to come, and literature describing them will take time to emerge. In contrast, the performance of the models that were part of CMIP3 is quite well known, and there is a wealth of literature describing them.

A potential solution was discussed, which takes advantage of the fact that the RCPs intentionally cover a wide range of forcings at the end of the 21st century. The SRES emissions scenarios that approximate the range of the RCPs which were also the drivers for the CMIP3 model runs, can be mapped onto the most similar RCP, including the stabilization RCP's. The model runs from the CMIP3 archive could then be scaled to RCP-based results, assuming that the relatively small differences in forcings are either unresolvable in climate system results or can be scaled statistically in some reasonable way. The details of the scaling would need to be worked out, but this approach would result both in using models in which the scientific community has some familiarity and in providing a way to compare results with the newest emerging literature, and with the fifth IPCC assessment. In addition, an overview of differences and similarities in expected future conditions based on the CMIP3 and CMIP5 approaches could be included in the 2013 Assessment report.

How will inter-sectoral impacts be modeled in the NCA?

Single sector impacts will continue to have an important place in the NCA, as they have in previous climate impact assessments. But one of the weaknesses of such studies, also identified in previous assessments, is that impact sectors clearly interact with one another. For example, water issues interact with energy, agriculture, and ecosystem issues. Issues of energy supply and demand compete for land with agriculture, forestry, and unmanaged ecosystems. Impacts on ecosystem processes and biological diversity interact with each other and with overall needs for water and for agricultural productivity. If the NCA is going to be able to characterize and evaluate both impacts and adaptation strategies, it is clearly important that we begin to understand how these impact sectors interact with each other.

A potential solution to this problem may have two components. One component could be to ensure that there are multi-sectoral, interdisciplinary teams that are tasked with analysis of such interdisciplinary, inter-sectoral problems. For example, within a region, a small team that includes hydrologists and agricultural experts might be tasked with the assessment of those cross-sectoral issues. A second component of a potential solution could be to begin using models that are explicitly multi-sectoral. Such models are beginning to appear in research groups around the country, i.e. groups in which models of

hydrology, agriculture, ecosystems, etc. have either been coupled directly, or in which groups of investigators in those disciplines have been assembled specifically to address interdisciplinary resource problems. A challenge to this second mode is that the integrated models are only beginning to appear in the scientific literature, so there would need to be careful coordination with sponsoring agencies and the scientists themselves to ensure that stakeholder consultation occurs and that only well-reviewed results are used in the NCA.

How might interdisciplinary, integrated assessment models be used in the NCA?

Of all the types of models that are relevant to the NCA, the integrated assessment models (IAMs) are perhaps at the most preliminary stage of development in terms of their use for impact and vulnerability analyses. Integrated assessment models have a long history of use for research into greenhouse gas emissions. Such models have been used for decades to investigate scenarios of the evolution of the energy system in terms of the mix of technologies for generating and using energy services and the resulting greenhouse gas emissions as a function of economic growth and development and changes in energy or climate policies.

However, as described in Janetos et al. (2009), the IAM community puts high priority on developing capabilities to use the IAMs to describe climate impacts and adaptation in a way that will allow quantification of tradeoffs and sectoral interactions. There is, in addition, a long history in the IAM community of driving individual resource models with output from the IAMs to understand regional impacts. Each of the current modeling groups has existing efforts underway with respect to land-use futures, biofuels, and both ecosystem and agricultural productivities that could be used in the context of the NCA.

Whether existing research efforts on regional integrated modeling will be immediately accessible by the NCA depends critically on timing of integrated assessment model development. If model results and published papers need to be completed and evaluated within the next 12 months, it will be very difficult to use the newest, most sophisticated models. This situation is analogous to the problems that are likely to be encountered in attempting to use CMIP5 global climate models.

How will model uncertainties be characterized and quantified in the NCA?

Workshop participants were of the opinion that a transparent treatment of model uncertainties in all types of models used in the NCA is highly desirable. This perspective was not limited to climate models — although that is where the topic has arguably received the most attention — but should also extend to the use of impact models in different domains (e.g., ecosystems, agriculture, and hydrology) and to the use of integrated assessment models. The NCA will likely want to make available all papers and documentation of model performance, model sensitivities to different parameters, evaluations of models compared to observational data, estimates of structural uncertainty (should they exist), etc. However, because it is unlikely that there will be new resources available for additional uncertainty characterization, the NCA will need to rely primarily on already existing studies for this information. Of particular concern is the development of methods for characterizing the uncertainties that are inherent in regional climate scenarios. The workshop reviewed several different methods for producing regional scenarios and methods for evaluating the uncertainties associated with the scenario products themselves. Uncertainties in regional model results should be of high priority because regional climate scenarios will be used extensively throughout the NCA.

It would be useful to have an overall guidance document on model uncertainties as reference material for the NCA, especially since much of the modeling itself may be used in a decision-making context rather than in a purely scientific context. So the degree to which different uncertainties in a model actually matter to a particular decision is likely to vary a great deal from case to case, and the NCA needs a clear way to systematically describe this character of model uncertainties.

Chapter 2:
Modeling and Scaling Issues in the National Climate Assessment

Anthony C. Janetos, Pacific Northwest National Laboratory; Kathy Hibbard, Pacific Northwest National Laboratory; Noah Diffenbaugh, Stanford University; Don Wuebbles, University of Illinois; Katharine Hayhoe, Texas Tech University; Katharine Jacobs, Office of Science and Technology Policy; Sheila O'Brien, U.S. Global Change Research Program; George Hurtt, University of Maryland; William Collins, Lawrence Berkeley National Laboratory

2.1 Introduction

The purpose of this chapter is to address a series of issues related to the use of models in the National Climate Assessment (NCA). While a major focus is on the use of climate models, there are also many other types of models that will be used in the NCA—ecosystem models, agricultural productivity models, integrated assessment models, hydrologic models, etc. But while distinct issues and assumptions are associated with each of these different kinds of models, there are general features of using models in the NCA that should be considered. This chapter begins a dialog on the use of models in the NCA and on longer-term research investments that are necessary to ensure that future Assessment activities can build on a continually improving foundation of models.

2.2 Background

The NCA 2013 report is required under the Global Change Research Act of 1990, which says in Section 106 that a "Scientific Assessment" must be prepared not less frequently than every four years and delivered to the President and Congress. This Assessment must

- Integrate, evaluate, and interpret the findings of the U.S. Global Change Research Program (USGCRP) and discuss the scientific uncertainties associated with such findings;
- Analyze the effects of global change on the natural environment, agriculture, energy production and use, land and water resources, transportation, human health and welfare, human social systems, and biological diversity; and
- Analyze current trends in global change, both human-induced and natural, and project major trends for the subsequent 25 to 100 years.

This last requirement to analyze future trends requires the use of models at various scales, and also the ability to build scenarios that help describe and analyze future conditions where changes in climate are only one of a myriad of changing conditions. Assessments serve important functions by providing the scientific underpinnings of informed policy. They also serve as progress reports by identifying advances in the underlying science, providing critical analysis of issues, highlighting key findings and key unknowns that can improve policy choices, and guiding decision making related to climate

change. The approach envisioned for this third National Climate Assessment is a comprehensive assessment of climate change, impacts, vulnerabilities, and response strategies within the context of how communities and the Nation as a whole can work to create sustainable and environmentally sound development paths.

A primary goal of the National Climate Assessment is to establish a permanent assessment capacity both inside and outside of the federal government. This new capacity will be an ongoing process that draws upon the work of stakeholders and scientists across the country to produce future assessments. Assessment activities will stimulate expanded capacity to

- Conduct research relevant to vulnerability to climate stressors;
- Observe and project impacts of climate change within regions and sectors;
- Develop consistent indicators of progress toward reducing vulnerability; and
- Allow for the production of a set of reports and Web-based products that are useful for making decisions at multiple levels.

Strategic planning for the new National Climate Assessment began in early 2010. The Interagency National Climate Assessment (INCA) Task Force, which includes members from all 13 agencies participating in the U.S. Global Change Research Program (USGCRP) and several others, produced a draft strategic plan. To ensure that the vision and implementation of the NCA are responsive, the Assessment will solicit input from a broad range of stakeholders, decision makers, and concerned citizens. A federal advisory committee will review research results and technical inputs prepared by interested organizations, groups, and individuals; oversee the Assessment process; and prepare the 2013 Assessment report.

2.3 Goals and Objectives of the NCA

The vision for the National Climate Assessment is to establish an ongoing, inclusive national process that
- Synthesizes relevant science and information;
- Increases understanding of what is known and not known;
- Identifies needs for information related to preparing for climate variability and change and reducing climate impacts and vulnerability;
- Evaluates progress of adaptation and mitigation activities;
- Informs science priorities;

- Builds assessment capacity in regions and sectors; and
- Builds societal understanding and skilled use of assessment findings.

Although the definition of regions to be used in 2013 NCA report is still under consideration, it has been noted that the ability to deploy information on the Web will significantly relieve pressure on defining regional boundaries. If the NCA can assemble information at a number of national, regional and local scales, the exact boundaries of the regions become less important. However, at a recent regional and sectoral workshop there appeared to be consensus that for the report there will be regions that are roughly analogous to those defined for the 2009 report, with adjustments to use state boundaries wherever possible. There is a strong desire for both understanding regional climatology and having the capacity to project regional conditions at multiple temporal scales, including seasonal to inter-annual, decadal, and 50–100 years. The need to understand change in both a transient and end-point framework was also noted.

2.4 The Role of Modeling and Associated Uncertainties in the NCA

The charge for the NCA, as laid out in the Global Change Research Act, requires an assessment of potential impacts and trends 25 to 100 years into the future. This charge challenges the modeling communities that will be involved in the NCA. First, it requires that both physical and economic models not only be used in a way that captures our current understanding of climate change, its impacts, and possible response strategies, but also to provide forward projections. Thus, there is an inescapable requirement to bring expertise from these disciplines into the Assessment process using a variety of models capable of projections. Second, the NCA will need to be completely transparent about the abilities of the various models used to do projections; where there are measurement, parameterization, or structural uncertainties; and the degree to which those uncertainties affect the confidence in findings. Lastly, this also highlights the fundamental need for the development and application of fully coupled models of human and natural systems.

The following sections discuss the potential roles of different families of models in the National Climate Assessment.

2.4.1 Climate Models

Uncertainty in modeling future climate change can be characterized broadly as resulting from three main sources: (1) internal or natural variability, characterized by model initial conditions and internal variability; (2) model structure and parameterization; and (3) alternative emissions trajectories that result from human decision making (Hawkins and Sutton, 2009). For both temperature and precipitation, analyses suggest internal variability is the dominant source of uncertainty over shorter time frames of years to decades. Model uncertainty dominates through mid-century for global temperature and end-of-century for regional temperature change and for precipitation at all scales, while scenario uncertainty becomes the main source of uncertainty in global temperature by the end of the century (Hawkins and Sutton, 2009; 2010).

While short-term global climate model experiments are being conducted at non-hydrostatic resolutions, most global model experiments are still confined to resolutions that do not capture the fine-scale spatial and temporal heterogeneity that can be critical for determining climate change impacts. Land surface features such as topography and land cover contribute to the response of local and regional climate to elevated greenhouse gas concentrations, and it is critical that assessments of the potential impacts of climate change consider physics at these fine scales. Capturing such physical details is, however, computationally demanding, particularly for the global domain.

Nested, or regional climate models offer capabilities for conducting targeted, high-resolution experiments with reduced computational demands. These nested experiments are, however, also computationally intensive, particularly given the many dimensions of the climate change uncertainty space that must be probed. For instance, high resolution is necessary to capture climate processes at the local scales that ultimately determine climate change impacts, while large domain extent is necessary to capture the influence of fine-scale processes on large-scale dynamics and thereby to accurately project local-scale changes across a large geographic area. However, few of these model experiments are dynamically nested to allow for simultaneous two-way exchange of information between the nested regional and global models.

Ensemble methods that generate multiple model realizations of a single experiment allow for quantification of the likelihood of simulated climate changes due to prescribed changes in climate forcing (e.g., atmospheric greenhouse gas concentrations) and uncertainty in climate sensitivity rather than internal variability. Likewise, long-term model integrations (e.g., centennial) make it possible to capture climate response to transient changes in radiative forcing, including potential changes in multi-annual and multi-decadal climate variability.

Additionally, relative to observations, the global climate model projections under-sample the allowed range for regional climate prediction. In other words, the significant range of unlikely, anomalous change represented through statistics (e.g., low-probability, high-impact events) are not contained in the probability functions implied by global climate models.

Finally, scenario uncertainty addressed by employing a range of emissions or concentration scenarios allows for consideration of the range of likely future human development pathways, while model intercomparisons and parameter sensitivity studies are needed to constrain the uncertainties associated with model formulation.

Taken together, these three sources of uncertainty demand a large suite of high-resolution model integrations covering a range of plausible model initial conditions, parameterizations, and resolutions, as well as alternative scenarios to investigate uncertainty associated with their underlying assumptions. While such integrations are becoming more common, computational and human resources remain limiting factors in achieving complete high-resolution ensemble calculations.

2.4.2 Impact Models

Using global climate model simulations as inputs, impact models have been used to evaluate and assess the consequences of changing climate for human and environmental systems. In contrast to the physically-based global climate models, impact models can simulate multiple aspects of biogeochemical, biophysical, rule-based, conceptual and integrated-response models through various techniques including process and mechanistic simulation, statistical and empirical relationships, and time series analyses.

Uncertainty in impact models arises primarily from model uncertainty—specifically, the degree to which the models accurately incorporate and are able to simulate the many factors influencing a given system or sector and how climate change might interact with existing drivers of change. Additional uncertainty arises from the fact that in the majority of impact studies, global climate model simulations typically are downscaled from a relatively coarse spatial resolution of 1° to 3° of latitude or longitude to a resolution that is relevant to the process or mechanism that the impact model is addressing, generally at local or regional scales, sometimes as fine as 100 m. Issues related to downscaling and regional climate projections are addressed in subsequent sections of this document.

Many different types of models have been used to assess climate impacts. Impact models include: ecosystem models, which calculate fluxes of materials and energy within and to and from ecosystems and generally are used to evaluate the ways in which such system variables such as net primary productivity respond to changes in climate; hydrologic models, which simulate water flow and quantity at a variety of spatial scales; dynamic vegetation models, which simulate disturbance and long-term trends in vegetation geographic distributions as a function of change in the climate system; and agricultural productivity models, which simulate changes in crop yields, water demand, and other agricultural inputs and outputs. These are just a few of the many types of models that have been used in past climate impact assessments and will have a role to play in the current NCA as well. As impacts that result from climate and or land use change can in turn affect future climate, it is becoming increasingly clear that capabilities for coupling predictions of impacts back into climate calculations are a key development priority.

Previous U.S. assessments (e.g., NAST, 2001) and the Climate Change Science Program Synthesis and Assessment Products provide information about how climate change could impact productivity, quality, and vulnerability within various sectors (e.g., forests, rangeland, water, livestock production, ecosystem, coastal, marine, human health, and transportation systems), with an acknowledgement of these limitations. As natural resources are constrained or limited through extraction or competition, it will be important that this NCA capture integrated analyses that incorporate simultaneous and multiple sector analyses (e.g., competition for land, water, and energy) at regional scales in the context of global change.

2.4.3 Integrated Assessment Models

Integrated assessment models (IAM) simulate the interactions between human decisions in the energy sector and both carbon and physical climate systems, as components of an overall Earth system approach. Historically, they have been used to simulate greenhouse gas emissions trajectories as a function of different development assumptions or mitigation targets, usually over the 21st century. IAMs include the major interactions between the human economy and other components of the Earth system, and most include a reduced form model that can be used to calculate global radiative forcing from human-produced greenhouse gases, if desired. All models include representations of agricultural productivity as needed to meet food demands, although these are generally not linked to climate changes except in the broadest sense.

Integrated assessment models differ significantly in the degree of foresight that they assume in their simulations, in the delineation of regions for analysis, and in the degree of specificity with which they describe energy technologies. Over the past decade, there have been significant changes in the evolution of IAMs, both in the U.S. and internationally. Although there are only a small number of sophisticated IAMs in existence, more are in the process of being developed. In creating these more sophisticated models, emphasis is being placed on improving model representations of processes in the terrestrial carbon cycle and their linkages to climate models. The most complex of these models now include interactive land-use components in order to enable carbon accounting. These models can also be linked to process-based ecosystem or agricultural productivity models; operate at multiple spatial and temporal scales; and can be used to generate probability density functions of emissions trajectories. Janetos et al. (2009) provide more detailed information about the research agenda that has been delineated by the IAM community.

A critical feature to note about IAMs is that although they are based on empirical observations on the energy and agricultural sectors and on the relationships between energy production and greenhouse gas emissions, they are not used to predict societal decisions. Rather, they are used to evaluate the potential consequences of different decision trajectories, usually with respect to mitigation targets. As they have developed, many models now also include land-use decisions as part of their analysis—at least that component of land-use that

is directly related to meeting demands for food and that interacts in various ways with the energy system (Reilly and Paltsev, 2009; Wise et al., 2009). These relationships are captured in the models' contributions to the IPCC fifth assessment process, for example, in their calculations of the emissions and greenhouse gas concentration trajectories known as representative concentration pathways (RCPs) (Moss et al., 2010; Thomson et al., 2011).

The primary source of uncertainty in IAMs arises from climate sensitivity and damage functions (an economic functional representation of negative impact of climate change) that are integral to the primary purpose of these models: that of representing the potential evolution of human society and economics. This uncertainty tends to interact with uncertainties associated with emissions projections, including aerosols such as sulfur. The largest overall sources of uncertainty in IAMs appear to be future demands for energy in the developing world as incomes and labor productivity increase and with technological changes in energy production (Scott et al., 1999).

In an analogous fashion to climate models, the integrated assessment models are evolving from their primary focus on global emissions to concerns about regional changes, and from mitigation to the other aspect of human decisions about climate challenges, adaptation. The major modeling groups either have now or soon will have regional versions of their IAMs, and these are being designed in ways to interact closely with more detailed models of various impact sectors, including impacts on the energy sector itself. And because both the global models and their newer regional versions deal explicitly with interactions among different sectors and Earth system components, they are potentially well positioned to address the questions of multiple stress and multiple sector responses that are important to the NCA. In the future, coupling these models to create fully functional climate, carbon, and land-use models will ultimately be needed for the most accurate predictions.

2.5 Establishing User Needs in the NCA

The NCA is being built around the concept that society needs the best available science in order to understand: (1) how the climate system is changing, (2) what the impacts are likely to be, (3) what uncertainties or risks are associated with these potential impacts, and (4) what additional information is required to build resilience into both natural and human systems through societal response and sound planning. Such information may take many different forms—from best estimates of regional climate change, to estimates of the costs and potential efficacy of different adaptation decisions, to estimates of the extent and cost of potential mitigation actions and their environmental consequences. Public and private decision makers, natural resource managers, city planners, and many others have needs and desires for information, much of which will ultimately come from models or from the use of models to extrapolate or interpolate available data. It is of high priority to understand and incorporate these models into the NCA in order to determine the degree to which the scientific modeling communities can best respond.

2.5.1 Regions

A strong focus in the new Assessment is on key sectors to define specific vulnerabilities and adaptation options. At the same time, people of the U.S. identify and make decisions within the regions in which they live, including their shared issues, resources, and landscapes. Thus, it is important to integrate the Assessment findings into a regional analysis.

In the past decade, since the first national assessment, there have been a number of local and regional climate impacts assessments, with foci ranging from specific cities like Chicago and New York to large regions like the Northeast and the Midwest. These assessments provided extensive input for identifying vulnerabilities and for considering adaptation and mitigation policy options, but because of time constraints, made limited use of downscaling approaches from global climate model results, generally using statistical approaches. For future assessments, including the 2013 NCA report, there is an opportunity to revisit what modeling and assessment capabilities are needed at the local and regional scale to enhance our ability to more accurately define the range of possible vulnerabilities and policy options at the local to regional scale.

Specifically, the current NCA should strive to effectively combine the results from a suite of regional climate models and from a limited set of very high-resolution simulations using one or more global climate models in conjunction with advanced statistical downscaling approaches to analyze the potential changes in climate change across the country. This approach will provide

essential input for vulnerability and adaptation analyses, representing regional variations in climate and climate change much more effectively than in previous assessments.

2.5.2 Impact Sectors

The first national assessment evaluated the potential impacts in the following five sectors: (1) agriculture, (2) forests, (3) health, (4) water, and (5) coastal areas and marine resources. The assessment used models to evaluate the potential impact of changes in the climate system on the overall status of each sector, and where possible, on the biogeochemical or physical processes inherent in each. The focus of the first assessment was primarily on using models to do this evaluation in concert with downscaled data from two general circulation models, both of which used the same business-as-usual emissions scenario (IPCC, 2001; NAST, 2001). Ensemble climate model results were not used, as ensembles at the time had not saved the right data to drive the sector models.

The second national assessment, *Global Climate Change Impacts in the United States* (Karl et al., 2009), used a very different approach. It began with a slightly larger set of seven impact sectors: (1) water resources, (2) energy supply and use, (3) transportation, (4) agriculture, (5) ecosystems, (6) human health, and (7) society. In addition, because the impacts report was largely a synthesis of the conclusions of the many Synthesis and Assessment Products developed by the Climate Change Science Program as well as literature published since the 2000 assessment, there was a much more mixed use of models at the sectoral level. In some cases, modeling played a key role in representing sector sensitivity.

For example, crop modeling continued to be an important feature of understanding agricultural impacts, as was hydrologic modeling for understanding the sensitivity of the water sector (CCSP, 2008). In other cases, e.g. transportation, modeling played a key role in evaluating the climate-related stresses to which the sector might be subjected. In that case, sea-level rise modeling in the Gulf Coast region was a critical feature of the assessment of the transportation sector's sensitivity to environmental stress, but modeling of the transportation sector itself was not as critical a feature (Savonis et al., 2008). And in other cases, e.g., ecosystems, the use of models at a sectoral level played a small role in the impacts report, as it focused primarily on observed changes in ecosystems that were related to changes in the physical climate system (CCSP, 2008).

As with the 2000 national assessment, no new climate modeling was done specifically for the 2009 report, but instead there was a reliance on the published literature to determine the possible impacts of lower and higher emissions futures. Unlike the first assessment, however, there were ensemble results available in the literature that could be used to characterize those potential outcomes, in part because the climate model output generated for the assessment was not used to drive impact models directly due to the more limited scope of the report.

In the current NCA, the list of sectors to be investigated stems primarily from existing legislation, but has been broadened substantially to include a series of inter-sectoral evaluations. For example, the interactions of the energy sector, water, and land use (including agriculture, forestry, and natural ecosystems) are critically important for understanding potential impacts of climate change on a regional basis (Janetos et al., 2009). The use of IAMs in concert with other models provides a set of tools for doing such analyses for at least some case-study sectors and regions of the U.S.

A primary challenge for the NCA will be defining the needs of the individual sector or multi-sectoral analyses with respect to modeling. Will modeling primarily be used to represent the changing physical and climate stresses on sectors of concern? Will sector models be used to evaluate sensitivities, or will there be some combination of these uses? What are the adaptation and mitigation decisions in each sector that would benefit from being informed by the use of models? And for extant models, what is the scientific community's ability to characterize the uncertainties in both data and model structure that affect the confidence that one should ascribe to the results of analyses?

2.6 Interaction of Adaptation and Mitigation

Until recently, mitigation by reducing greenhouse gas emissions or increasing their sinks as well as adaptation by altering behavior or management to address the consequences of climate change have been considered separate societal responses to climate change. But it is increasingly clear that both mitigation and adaptation strategies are necessary

and that those responses have the potential to interact with each other. A clear understanding of the trade-offs and consequences of these interactions as they apply to sectors and regions is needed.

There is a spectrum of characteristics and opportunities for mitigation and adaptation strategies. For example, physical (e.g., the availability of water or sufficient soil fertility) or economic (e.g., the availability of physical infrastructure) constraints towards the implementation of different energy technologies (e.g., biofuels) or mitigation strategies (e.g., carbon capture and storage) are challenges that could be considered jointly, rather than from sectoral (e.g., ecosystem versus energy) perspectives. Indeed, mitigation and adaptation strategies should be considered as a continuum of options, from the active policy to reduce greenhouse gas sources or enhance greenhouse gas sinks, to the modification of energy infrastructure, to the development of societal and natural systems resilient to change and the adaptive management of ecosystems. Some examples include the de-carbonization of energy sources and proliferation of electric vehicles for transportation on the mitigation side, and developing wildlife corridors, improving building energy efficiencies, and building sea walls on the adaptation side.

There likely are several options where mitigation and adaptation strategies can be mutually beneficial. For example, health experts have highlighted the potential co-benefits of designing cities that are more conducive to walking, which could both reduce greenhouse gas emissions from transportation and improve human health, thereby reducing vulnerability to climate-related stresses such as heat waves. Likewise, no-till agriculture often enhances the greenhouse gas sink and improves water use efficiency, thereby providing potential adaptation to decreased water availability. Further, enhancing forest health could reduce the number of forest fires and hence the associated carbon dioxide emissions, while also increasing forest resilience to drought.

Conversely, it is also important to consider potentially antagonistic relationships between mitigation and adaptation activities. For example, geo-engineering through stratospheric injection of sulfur dioxide could impact human health, agriculture, food security, and ecosystems through alterations to the hydrologic cycle and availability of photosynthetically active radiation for plant production. Likewise, the impact of land-based carbon sequestration on local and regional climate and hydrologic balance could be as large as the impact of elevated greenhouse gas concentrations.

Specific information needs for considering integrative mitigation and adaptation strategies include an understanding of
- Future climate change and associated impacts,
- Adaptation and mitigation options (including interactions),
- Potential policy decisions (interactions of national, state, and local),
- Levels of confidence and uncertainties, and
- Options for continued stakeholder interactions.

2.7 Right-Scaling Model Simulations

The challenge of effectively using models developed for one class of problems that have a particular spatial and temporal dimension and applying them to other problems is at the heart of the Assessment process. Typically, this has been discussed most often as the challenge of downscaling climate model output to spatial or temporal scales that are useful for analyzing potential climate impacts, but the problem is not limited to the use of climate models. For example, some model intercomparison studies (e.g., Melillo et al., 1995), used model-based interpolations of historical and current climate observations to spin-up ecosystem models, and then downscaled global climate model output for future simulations. This required substantial thought and methodological development not only in dealing with model output, but also in the assembly of observations (which under-sample high elevations, for example) and consideration of how those data sets could be sensibly combined for ecosystem analyses.

The ecosystem models themselves are typically analogous to single-column models, and their implementation on larger geographic grids has often involved highly-parallel simulations. However, the incorporation of more realistic hydrology, including river flows and impoundments, creates a need for more sophisticated approaches to these simulations, as water balance and flow must be accounted for on a broader regional basis. Correct scaling in this context raises different concerns than in the climate modeling case. Hydrologic models tend to be in-between—some are intrinsically column models, while others are explicitly constructed for regional simulations with somewhat arbitrary areas. While in practice many of these models are implemented at 1° latitude × 1° longitude or 0.5° latitude × 0.5° longitude spatial resolution, these spatial resolu-

tions do not resolve underlying climate and soil heterogeneities, which vary at much finer scales, leading to potential errors in prediction. State-of-the-art models are pursuing 1 km and finer spatial resolutions with corresponding data requirements at these scales. To drive mechanistic models of plant photosynthesis and energy-water balance in these models, high temporal resolution (diurnal or better) is also required.

Integrated assessment models, in one sense, have both of these scaling issues to consider, but another as well. Economic and technological information needed to represent the energy sector is not collected on a gridded basis, but rather according to administrative or legal boundaries and by economic sector and is thus intrinsically not geographically specific. In some cases, geo-referencing such information is simply not reasonable. The challenge then becomes how to use that information in model frameworks that are evolving from investigating century-long strategic issues in global or large regional mitigation options to analyses of smaller regions and how to combine that information with gridded, physical data or model output.

2.7.1 Right-Scaling Climate Models

Local and regional features modify the influence of global climate change in ways that challenge even existing regional models. Simulating these patterns of change is essential to quantifying regional- to local-scale climate change impacts on both human and natural systems. Downscaling or simulation of climate variables at scales finer than those resolved by the global models therefore represents a fundamental tool for the ongoing Assessment process. Both dynamical (regional modeling) and statistical downscaling capabilities will be important to the NCA process.

2.7.1.1 Regional climate models

Akin to global climate models, regional climate models (RCMs) explicitly solve a set of equations representing the processes affecting climate, but at a much finer spatial scale. Regional model simulations, or dynamical downscaling, provide a full set of consistent climate variables including simulation of known feedback processes that may affect the stationarity of the relationship between large-scale circulation patterns and local climate. Regional climate models are currently limited to a spatial resolution of about 10-50 km (100 km²). Although some have a finer resolution of 3–4 km (9–16 km²) over very small regions, the models largely have not been tested and refined for use down to resolutions below this threshold.

The primary disadvantage of RCM simulations is that they are computationally demanding and, at this point, only a limited set of global climate model outputs are available at the high temporal resolution required to drive RCM boundary conditions. It is also important to note that RCMs are analogous to a digital signal, in that they represent only the subset of local features and physical processes that are explicitly resolved by the model.

For the NCA, it will be important to evaluate the capabilities of existing regional models, particularly with respect to their ability to represent climate processes and changes across the U.S. In terms of future development, finer spatial resolutions are needed, particularly over regions of high topographic variability, as are improved parameterizations of sub-grid-scale physical processes such as convection and turbulence.

2.7.1.2 Statistical downscaling models

There are many different statistical downscaling models, ranging from simple to complex. Most are adequate for average temperature changes at seasonal to annual scales, but there are large differences in their abilities to simulate daily values and particularly extremes. Precipitation presents additional complications, particularly related to determining the best choice and use of large-scale predictors.

In contrast to process-based models, statistical downscaling models operate much like an analog signal, incorporating all local influences on observations (including observational error) into future projections. Statistical downscaling methods are generally cost and time efficient; they can be generated at the scale of any observational data set, including station-based and gridded; they are easily transferable to a variety of applications and locations; and they directly incorporate real historical observations into future climate projections for a given location.

The primary limitation of statistical downscaling is the assumption of stationarity in the predictor and predicted value relationship. In other words, statistical models must assume little to no change in the climate system feedback mechanisms through time that connect large-scale circulation features to local-scale climate. Several studies have developed methodologies to test this assumption, comparing future regional model simulations to global climate model biases and statistical methods (Liang et al., 2008; Vrac et al., 2007). Initial results suggest that, at least in some locations, the assumption of stationarity holds, particularly under lower (as compared to higher) amounts of change.

Both statistical and dynamical downscaling models often produce projections of regional and local climate that are significantly modified from the global models, but both approaches can be affected by the quality of the global climate model results used in their analyses. Neither method is dynamically coupled with global models, meaning that they are not yet able to capture local- to regional-scale climate change feedback processes.

It is clear that any and all downscaling tools proposed for use in the Assessment process should be thoroughly evaluated using a standardized approach before being used in the Assessment. Despite the essential role of downscaling in regional assessments, there is no standard approach to evaluating various downscaling methods. Hence, impact communities often have only a vague awareness of limitations and uncertainties associated with downscaled projections. A standardized framework of physical and statistical tests should be used for evaluating and comparing downscaling approaches.

Initial application of such a framework (Hayhoe, 2010) to a broad range of downscaling methods and locations has identified the downscaling method used as a more important determinant of data quality than station location or global climate model for many (although not all) aspects of temperature and precipitation distributions. Key differences between downscaling methods arise particularly at the tails of the distribution. These differences can lead to projected changes in extreme heat days, for example, that vary by nearly an order of magnitude between various downscaling approaches applied to a given location.

The selection of regional models and statistical downscaling approaches is a key assessment task. There is no single best way to generate the high-resolution projections needed for assessing climate vulnerabilities and adaptation responses across the entire country. Both statistical and dynamical (regional modeling) methods have their unique strengths and limitations, and both approaches, especially in combination, can be powerful tools for assessment purposes. Whether derived using regional or global models, projected future changes should be averaged over climatological periods (20–30 years) to avoid overemphasizing short-lived trends or misinterpreting results suggesting that specific events may occur in specific years.

2.7.2 Right-Scaling Integrated Assessment and Other Models

Like global climate models, integrated assessment models (IAM) historically have been formulated to investigate problems of global emissions trajectories and the mix of energy technologies needed over very long periods of time (roughly a century) with long time steps (e.g., 5–15 years), without fine-scale, geopolitical specificity but with the differentiation of regions into large political domains (e.g., Western Europe or Annex I Nations). However, while these spatial and temporal frames are initially appropriate for the original problem set, the desire to understand the interactions with physical and ecological systems and to represent finer scales of implementation and the interaction of different sectors also demands that the IAMs be reformulated to address finer spatial scales and shorter temporal scales. Linking IAMs to physical and ecological models in more sophisticated ways is beginning to address some of these issues. Other issues are being addressed by reformulating the models to be truly regional in nature. This is one of the major research foci of the IAM research community, and the timing of results will need to be monitored carefully for the purposes of the National Climate Assessment.

2.8 Progress Over the Past 5–10 Years

2.8.1 Climate Models

Over the last decade, concerted efforts in the climate modeling community have focused on: (1) understanding and better quantifying the uncertainties inherent in model simulations of climate and climate change; and (2) improving model resolution and representation of physical processes important to the climate system.

To understand and better quantify uncertainty, multi-model ensembles are used to identify common features in projections of climate change. Models (CMIP3 and more recently CMIP5) have established formalized structures that enable models and evaluations against the climate record of the recent past. These efforts have provided an unprecedented resource for evaluating and comparing climate model simulations of the recent past and

projections for the near future, with over 500 articles based upon data sets produced and assembled by the Program for Climate Model Diagnosis and Intercomparison (PCMDI).

For the fifth assessment by the IPCC, PCMDI is serving as one of the three central archives for model results from the global climate modeling community. The model simulations have been commissioned under the auspices of CMIP5. The new elements of this intercomparison include a major focus on near-term, decadal-length projections designed for regional climate change and on predictions from the new class of Earth system models that treat the coupled evolution of physical, chemical, and biogeochemical climate processes.

2.8.2 Integrated Assessment and Impact Models

The historical role for IAMs has been to provide data and models relevant to understanding the scale and timing of the drivers of climate change over decades to century time scales. Over time, IAMs developed the capacity to simulate emissions and their associated greenhouse gases, by first introducing carbon dioxide, then simple climate, followed by multi-gas and finally, agricultural productivity and land-use decision making based on economic rules.

As mentioned earlier, impact models employ a broad array of process, empirical, conceptual and statistical tools. They are generally developed and applied to understand specific natural and human responses to climate change from primarily sectoral (e.g., agriculture, ecosystem, human health) perspectives. As such, many have evolved to consider finer scale process information or interactions with social dynamics in either off-line or simple ways. For instance, the most advanced terrestrial models have developed sophisticated representations of three-dimensional vegetation structures and calculations of carbon and nitrogen dynamics to account for land-use and land-management practices (e.g., grazing, fire suppression, forest harvest, and fertilization) and natural disturbances (e.g., non-anthropogenic fire). Additionally, some integrated models use matrices of age-states to account for cohort dynamics following natural disturbances such as forest harvest, afforestation, or forest fires. Abrupt or catastrophic changes in natural systems (e.g., wind throw events, insect infestations is forests, or changes in river routing) are not generally reflected in impact models; nor do these models account for water management (e.g., irrigation to energy infrastructure demands). However, active research in this area has led to advances in simulations of fire and the effects of tropical cyclones.

The challenge for integrated assessment and impact modeling communities is to develop an integrative philosophy that considers rapid changes or even systemic changes as a function of multiple interacting factors including climate change, harvest, environmental services and socioeconomic values such that impact responses reflect the diversity of the systems.

Climate, integrated assessment, and impact models are functionally linked through a range of shared interdependencies that include climate information (e.g., temperature and precipitation), atmospheric composition (e.g., carbon dioxide and other gases), land use, and land management. The need for consistency and linkages across these models has motivated the international community to develop a harmonized set of land-use scenarios that smoothly connects historical reconstructions of land-use changes (agricultural land-use changes and wood harvest), based on satellite and other data, with future projections based on IAMs in the format required by global climate models and Earth system models. These harmonized data products provide the first consistent set of land-use change and emissions scenarios for studies of human impacts on the past and future global carbon-climate system. The application of these harmonized data sets will allow for improved climate, ecosystem, and biodiversity predictions and impact assessments.

2.9 Currently Available Resources

2.9.1 Climate Models and Outputs

Simulations using CMIP3-generation global climate models and based on SRES scenarios represent the largest collection of consistent climate model results currently available. For the fifth IPCC assessment, the next generation of global climate models is being used to simulate responses consistent with the new representative concentration pathways for both centennial and decadal predictions. These new generation models incorporate a number of key improvements including higher resolution (typically 1° latitude × 1° longitude), the carbon cycle coupled with the climate system, as well as enhanced treatments of important physical processes including the cryosphere, oceans, clouds, aerosols, and land use,

all of which are known to strongly affect regional climate change and feedback processes.

Initial comparisons between CMIP3 and CMIP5 generation models suggest noticeable improvements in the ability to simulate regional precipitation patterns and seasonal temperature change, as well as important differences in the geographic distribution of projected future change. The reliance of both regional modeling and statistical downscaling on global climate model outputs as boundary conditions suggest there is a clear benefit to relying on simulations by the newest generation of global models. Higher-quality global simulations nearly always translate into higher-quality regional simulations.

Because of constraints imposed by the schedule for completion of the 2013 NCA report, reliance on results from the existing array of CMIP3 model results may be most feasible. Results from CMIP5 should be evaluated as soon as the new historical total forcing simulations are available, using criteria directly relevant to the simulation of regional climate variability and change over the U.S. To the extent possible, vulnerability analyses and downscaling studies should rely primarily on simulations from the latest version of well-established, well-documented global climate models that will be considered by the fifth IPCC assessment and that demonstrate adequate performance in both CMIP3 and CMIP5 comparisons and against the regionally-relevant criteria recommended above. High-resolution global climate model simulations over limited time periods can be incorporated to further enhance understanding of the veracity of global climate model simulations of regional attributes.

As downscaling and impact modeling can be computationally demanding, it will be useful to evaluate what minimum subset of global model ensemble calculations approach the multi-model ensemble means for required forcing to impact models. This result would enable the development of criteria to identify a subset of global climate models to be used for in-depth analyses (at least four to five models, but not the entire suite of CMIP3 and CMIP5 global climate models).

2.9.2 Integrated Assessment and Sector Models

Contributions from U.S. and international integrated assessment modeling teams to the representative concentration pathways (RCP) for the fifth IPCC assessment process are already available. These include both emissions and concentrations trajectories from the present day through 2100, with end points defined as specific values of radiative forcing. Of particular interest to the NCA are emissions, industrial activity, and land-use trajectories that have been downscaled to a regular 0.5° latitude × 0.5° longitude global grid. Because these simulations are complete according to radiative forcing endpoints for the global climate model experiments, there is no guarantee that the underlying socioeconomic assumptions for the published scenarios will be preferred for either sectoral or regional analyses desired by stakeholders in the NCA. Nevertheless, they constitute a set of consistent simulations of changes in energy supply and land uses that provide considerable comparability with literature assessed by the fifth IPCC assessment.

Because the downscaling procedures for emissions and concentrations from the IAM models have been developed for the RCPs for the fifth assessment, in principle they might also be applied for new model runs from IAMs. Which model analyses would make sense to do in the context of the NCA is not yet clear.

Articles describing hydrologic, ecosystem, dynamic vegetation, sea-level rise, and many other types of models used for impact studies already constitute a rich literature. Whether or not existing studies have resulted in archived data sets that are easily accessible and whether the climate or socioeconomic data used to drive the models will be consistent with the choices made in the NCA is unknown, but could be explored on a case-by-case basis. In most research and operational teams, the ability to do new model studies is not so much limited by access to computational resources as it is limited by financial support and available time. These pragmatic issues will need to be addressed by the NCA.

2.10 Characterizing Uncertainties

The challenge of characterizing uncertainties in models, scenarios, and our underlying knowledge about climate change issues has been analyzed extensively over the past decade (Hawkins and Sutton, 2009; 2010; Morgan et al., 2005; Moss and Schneider, 2000; Parson et al., 2003). For the most part, this literature has focused on the serious challenge of characterizing the uncertainties in the underlying science, as captured by models. This aspect of uncertainty is critical—the assumptions and attributes underlying models used to assess changes in the

climate system or used to assess impacts is clearly a foundation on which any assessment is built. There is, however, another aspect of uncertainty that is particularly important for assessments such as the NCA that are also meant to be useful in informing a variety of decision-making processes. This is the uncertainty in climate or impact variables that is the most important from the standpoint of those actually making decisions. This uncertainty has received less attention in the climate change scientific literature. But there is broad recognition that from the standpoint of making decisions about adaptation, management of natural resources, or changes in energy supply technologies, that there are many features of such decisions that are inherently uncertain but that are not necessarily related to the physical climate system. Alternatively, it could be the case that some, but not all, aspects of changes in the physical climate system are relevant to particular decisions. Thus, the goal of being relevant to decision makers introduces substantial complexity into the problem of characterizing uncertainties, compared to only considering the underlying physical science.

2.10.1 Evaluating Climate and Impact Models

There is a long and vigorous history of climate model evaluation, and such efforts are being enhanced further and formalized as part of the IPCC fifth assessment process. However, most of this effort has focused on evaluating the ability of climate models to capture purely physical climate metrics. While many of these metrics are clearly relevant for climate change impacts, methods for evaluating the ability of climate models to simulate climate impacts indicators have not been explored thoroughly. Just as climate model performance can vary substantially depending on the physical metric in question, it is likely that model evaluations could vary similarly between traditional physical climate indicators and impacts-oriented indicators. A new impacts-oriented paradigm may therefore be needed for reliable evaluation of the ability of different climate models to simulate climate change impacts.

A key challenge for climate model evaluation is developing methods for testing the ability of models to capture not only the baseline climate state but also the response of climate to changes in forcing. The past several years have seen a number of notable innovations in appraising the physical climate response, such as using the seasonal cycle to evaluate the representation of snow-albedo feedbacks and using inter-annual variations to evaluate the response of heavy precipitation to warm sea surface temperatures. Climate change impact assessments will benefit greatly from the development of similar methods for evaluating the ability of climate models to capture not only the relationship with baseline climate, but also the response to changes in climate.

One of the main challenges to the model evaluation enterprise is the absence of a unifying principle for the necessary and sufficient conditions a model must satisfy in order to provide reliable projections of the future. In the absence of such a principle, the community has developed a three-pronged approach to model evaluation using a combination of theory and empiricism. First, the processes that can be formulated from fundamental physical science are tested against benchmark calculations derived from the same basic theory. These processes include atmospheric fluid dynamics, radiative transfer, and reactive gas chemistry. Second, the predicted states of the climate system that emerge collectively from each component model are tested against a rapidly expanding suite of in situ and remotely sensed observations at local, regional, and global scales. The tests also include comparisons against integrative and comprehensive analyses of the historical climate record, for example global reanalyses underway in support of the Year of Tropical Convection Project. Third, the features of the models most directly relevant to impacts and adaptation are examined using more specialized statistical tests designed to extract significant signals from the underlying, unforced variability of the climate system. The detection and attribution of enhanced risk of climate extremes is an emerging and increasingly important activity in this field.

Two important trends in model evaluation have emerged over the last decade. First, model developers are increasingly aware that in-sample tests—using the same data set for model construction and parameter estimation and for model evaluation—are insufficiently stringent tests of climate models. This has led to an increasing emphasis on severe evaluation tests analogous to closure studies from observational field campaigns. In these severe tests, the main processes governing a climate property of interest are each evaluated against an independent data set.

For example, the climate community traditionally evaluated the fidelity of cloud radiative effects using the same satellite data sets used for ad hoc adjustments to the modeled top-of-atmosphere energy budget required for drift-free climate simulation. The community now evaluates radiative effects using independent and rigorous measures of cloud physical and microphysical properties from surface- and space-based radar (e.g., CloudSat), lidar (e.g., CALIPSO), spectral retrievals (e.g., MODIS), and in situ measurements. This evaluation process helps ensure that the empirical fidelity of each climate phenomenon of interest is based upon accurate representation and simulation of its underlying physical processes.

The climate community is beginning to use quantitative and repeatable measures of climate performance based upon various metrics or scores. Much of the evaluation activity has historically relied upon quasi-quantitative comparisons of modeled and observed fields based upon maps or transects through the atmosphere and ocean. While useful for model development, this type of analysis has some limitations. The analysis approach is

- Neither strictly repeatable nor purely quantitative, so two observers may draw different conclusions (even with respect to sign) from visual inspection of a given map;
- Difficult to extend to a multi-model ensemble given its optimization for pairwise comparisons; and
- Difficult to extend to evaluation of a time series of model versions for the same reason.

In order to address these deficiencies, the community has increasingly emphasized the use of collections of quantitative scalar measures of model error relative to observations, benchmark process calculations, and reanalyses of the climate record. This trend has been marked by the rapid introduction of several techniques, including Taylor diagrams (Taylor, 2001) that highlight errors in the correlation and variance of simulated climate patterns. Another technique is based upon a basket of numerical scores for model performance developed under the ClimatePrediction.net project for evaluating ultra-large ensembles of climate models. This technique has been adopted widely for some of the CMIP3 multi-model ensemble and will likely be applied to the CMIP5 ensemble as well. While these new methods provide several distinct advantages relative to traditional evaluation methods, there is still no underlying theory for which measures and metrics of model error must be optimized to provide robust and reliable projections of future climate change.

Several methods have been widely adopted to complement these trends:

- **Instrument Emulators.** In the past, in order to facilitate the comparison process, measurements were post-processed to retrieve quantities predicted or diagnosed by climate models. The current practice is to simulate the measurement systems and measurements directly from the model state. One of the early instrument emulators developed for the Cloud Forcing Model Intercomparison Project (CFMIP) reproduced the International Satellite Cloud Climatology Project (ISCCP) cloud data set. New emulators have been developed for the MODIS, AIRS, CloudSat, and CALIPSO instruments (among others). These emulators eliminate the assumptions implicit in retrieval processes and reproduce instrument spatial and temporal sampling for more reliable comparisons.

- **Initial Value Experiments.** One way to test fast climate processes forced with realistic meteorological conditions is to operate the climate models as numerical weather prediction systems. In this mode, the fidelity of a time series of short-term projections is evaluated against a coincident set of observational data. The dual advantages of this approach are minimization of background state error and elimination of confounding feedbacks between the process of interest and the test of the climate system.

- **Paleoclimate Experiments.** The community increasingly is relying on out-of-sample tests using the paleoclimate record to test models under climate regimes very different from present-day conditions. Recent successes in this arena include the simulation of 20,000 years of the Holocene using the Community Climate System Model (Ammann et al., 2007).

- **Comprehensive Regional Model Intercomparisons.** These include the European Prudence and Ensembles and the U.S. initiated North American Regional Climate Change Assessment Program. These projects have yielded important information on the relative contributions to the uncertainty in regional-model error from for example, model resolution, model physics, and the realism of the driving boundary conditions.

- **Transferability Studies.** It is increasingly common to test regional models of a phenomenon (e.g., the Southwestern U.S. monsoon) on other analogous regional phenomena to investigate the generalizability of the model formulation.

2.10.2 Incorporating User Needs

Incorporating the information needs of different user communities in assessment activities, and then using those needs to decide which aspects of uncertainty are the most important to decision making has been attempted far less frequently than a scientific characterization of model uncertainties. For example, in the 2000 national assessment the major attempts in this direction were to try to gather input from a broad spectrum of potential stakeholders in different parts of the country and associated with different sectors. Twenty-one workshops were held in different parts of the U.S. with the expressed purpose of attempting to elucidate the issues of most importance to stakeholders in each region irrespective of climate change. And while that information did lead to exploration of some issues in more depth than others, there was not an overall analysis of which elements of the physical climate system or of physical impacts were the most important in determining decision types. This may be in part because even that assessment, which included a very large and explicit attempt to involve many stakeholders, was not specific about what decisions or what types of decisions it was meant to inform. The NCA is not yet at a stage of development where actual decisions can be specified for which information is desired.

2.11 Characterizing Uncertainty in Climate Scenarios

Like the Earth system models and global climate models used to simulate climate, integrated assessment models (IAM) are subject to their own set of uncertainties and assumptions that propagate into scenarios, which are the basis for many climate change simulations. In the most reductionist sense, the IAMs consist of component models for the energy market, the carbon cycle on land surfaces, and reduced or simplified representations of the climate system. The energy market models project emissions of anthropogenic radiatively active species as well as land-use and land-cover change that interact with the other components. The energy models are subject to uncertainties from a variety of sources, including external boundary conditions (e.g., projections of future population growth), idealizations regarding the operation of the markets, simplifications regarding the environmental effects of energy consumption, and factors that compound over time such as the future rate of enhanced agricultural productivity. Another obvious source of uncertainty stems from abrupt market transitions that are inherently unpredictable from the historical record (e.g., introduction of market-disruptive technologies).

The interactions from the other components back onto the energy market may also represent major sources of uncertainty. The magnitude of this uncertainty depends upon the nature of the climate mitigation policy targeted by a given scenario, for example limitations on climate forcing imposed in a given scenario. The reason is that many policy targets implicitly represent the sum of a forcing contributed by anthropogenic emissions and a radiative feedback from natural sources and sinks in response to climate change. Uncertainties in the natural feedbacks, combined with a fixed target for future forcing thereby automatically introduce uncertainties in the anthropogenic emissions. In the third assessment by the IPCC, the primary SRES scenarios were generated using multiple carbon cycle models forced with climate states from global climate models with low, intermediate, and high climate sensitivity. The magnitudes of the uncertainties stemming from feedback effects are evident in the variations among the well-mixed greenhouse gas trajectories for multiple realizations of a single marker SRES scenario, for example the A1B scenario.

Other significant issues are related to different and potentially inconsistent formulations of the carbon cycle and other climate system processes in the IAMs and Earth system models (ESM). In order to enable the exploration of a large range of future scenarios, IAMs typically and by necessity employ a simpler representation of the carbon cycle and other climate processes than do the ESMs. The IAM and ESM communities also devote their development efforts to their respective topical domains with the result that the carbon and climate processes within the IAMs may sometimes lag the latest advances in physical and biogeochemical climate science. While one obvious solution is to integrate the energy market mechanisms in IAMs into an ESM, this would present two disadvantages: (1) the computational speed of the IAM would be lost due to the overwhelming computational demands of the host ESM; and (2) the resulting scenarios would become contingent on the particular representation of carbon and climate processes in a single ESM. A more promising solution would be to maintain dual-use IAMs that could operate as stand-alone systems or as energy markets integrated with ESMs, and to develop flexible and extensible community-wide standards for coupling IAM and ESM components

to enable experimentation across IAM and ESM development teams.

2.12 Implications for Future Research

Each of the modeling topics discussed is itself a subject for further research. While there is a short-term aspect to the current NCA, for an ongoing Assessment process, there is clearly a need for further research in each of these topics. Topics, which in our judgment are amenable to making progress in the mid-term or that strengthen the foundation for longer-term progress, include:

- Continued development of methods to evaluate the scientific uncertainties in each family of models used in assessments: climate models, impact models, and integrated assessment models;
- Development of specific methods for evaluating the inherent uncertainties in climate downscaling methods;
- Development of uncertainty methodologies associated with decision making;
- Development and testing of models and development of analyses that incorporate multiple sectors (e.g., energy, water, and land use) simultaneously.
- State-of-the-art terrestrial models incorporating very high-resolution climate data (at least 1 km and hourly) to avoid process errors from using overly averaged input data.
- Development, testing, and application of coupled climate, socioeconomic, and ecosystem models to address strong interdependencies. At the intersection of these efforts is the need to continue to develop and integrate advanced high spatial resolution models of land use and land management.
- Robust multi-scale, multi-variable observation system to support modeling and analyses. The strategic development of such a monitoring system is a high priority for model validation and testing.

Chapter 3:
Workshop Report

3.1 Introduction

The principal goal of the Modeling and Downscaling Workshop was to evaluate the current status of science, tools, and capabilities for developing, evaluating, and using climate, integrated assessment, and other applicable models for use in the U.S. National Climate Assessment (NCA). The workshop agenda included targeted discussions of issues associated with downscaling of various modeling results. Other topics discussed included (1) major model types, applications, strengths, and limitations; (2) model inter-dependencies; (3) scaling and multi-scale interactions; (4) matching models to user needs; (5) models and ensemble methods; (6) uncertainty and its implications; (7) data quality; (8) scenarios and critical assumptions; and (9) insights about developing a long-term, sustainable Assessment process.

The intent of the workshop was to consider these topics so as to identify compelling and useful resources for NCA teams as well as opportunities within current funding constraints to leverage existing modeling and scaling efforts or results. In addition, participants provided ideas, for structuring NCA modeling efforts to address the needs of a broad range of users and stakeholders, while also targeting the primary needs of assessment groups. The intent was to treat traditional methodological issues adequately while addressing new challenges and taking advantage of innovations.

Sixty-nine people, both federal and non-federal, attended the workshop, including approximately a dozen who also participated in a two and one-half day workshop on Scenarios for Assessing Our Climate Future: Issues and Methodological Perspectives for the U.S. National Climate Assessment, immediately preceding this meeting. The synergies between these two conversations proved to be important. Participants in the modeling workshop represented a broad cross section of experience with a variety of modeling techniques at multiple scales, including global climate models, statistical and dynamical downscaling, sectoral climate impact models, integrated assessment models, etc. Among the speakers were several people who have been engaged in modeling efforts in the context of previous NCA and Intergovernmental Panel on Climate Change (IPCC) activities, people involved in translating climate information for specific decision support activities, and a number of people with regional and sectoral interests who are users of such

information. The workshop occurred December 8–10, 2010 in Arlington, Virginia. The U.S. Department of Energy generously supported the meeting.

This report summarizes diverse ideas and information produced by the workshop, a number of comments by individuals, and recommendations as a resource for the NCA Development and Advisory Committee (NCADAC). This report is not a complete summary of the workshop, but provides an overview of the points that were emphasized throughout the meeting. The agenda for the meeting can be found in Appendix A, and Appendix B lists participants.

3.2 Summary of Plenary Presentations and Discussions

A series of workshop presentations provided background information about the use of models in national climate assessments. Several presentations addressed the use of statistical processing and model analyses to derive data at spatial and temporal scales relevant to the National Climate Assessment. Additional topics included approaches for characterizing uncertainty, interfaces between model analyses and users of the results, and requirements for model results addressing specific sectors. The following sections summarize major points of workshop presentations and associated discussions.

3.2.1 Introduction to Climate Modeling and Modeling Approaches for the National Climate Assessment

Although climate models are often used for predictive purposes, they do not represent all aspects of the Earth system that affect climate and are never completely accurate. However, if used properly, models can be powerful tools for understanding how the climate system functions now and can provide an important means of exploring a range of possible future conditions. To the extent that numerical models represent our understanding of the components and processes of climate and connected Earth system effects, they can be used to help characterize the attributes of components and processes that are incompletely observed by measurements. Chapter 2 of this report includes an extensive discussion of types of models and their applications at various scales.

With respect to increasing documentation of global impacts and better understanding of the mechanisms underlying those impacts, climate change perspectives are evolving quickly. While climate change issues have been considered primarily at global and national scales, there is now significantly more research at regional and local scales. In the past, modeling studies projected climate change over a century. Increasingly, studies focus on much shorter periods including decades. There is also a much stronger relationship between scientific research and policy making, with more emphasis on providing scientific foundations for decision-making processes. In addition, substantial research efforts now focus on the intersection of mitigation, adaptation, impacts, and vulnerability rather than on studies of these issues as independent topics. More attention is being given to integrated studies of relationships between sectors and multiple stressors. All of these developments result from our increasing understanding of physical and biogeochemical processes, increasing computational resources, and more sophisticated efforts to understand the complexity and inter-dependencies of physical, ecological, and human systems.

There is enormous complexity within the physical, chemical, ecological, and human systems being modeled. Even where components of the Earth system are well understood, computational resources restrict model accuracy and veracity and thus our ability to address important questions. With respect to contending with complexity in the Earth system, important science questions include

- How do we evaluate model capabilities for capturing climate responses? Are we getting the right answers for the wrong reasons?
- Are uncertainty evaluations different from the impacts perspective than in terms of physical climate change?
- How do we distinguish differences between models from uncertainty associated with variability in the Earth system?
- How do we quantify the limits of predictability inherent in the Earth system? Limits on predictability may not be the same as the spread in model simulations, even if ensembles of model simulations allow us to create probability distribution functions.
- How do we communicate unknowns and what cannot be known in an effective manner, even in cases where ensembles of model simulations are used to derive probability distribution functions?

The National Climate Assessment is confronted by a number of options with regard to models and model results. There is a need to distinguish between

- Modeling and model results in a perfect world versus what is reasonable given time and resource constraints;
- Models that incorporate the latest disciplinary research versus use of a well-vetted set of existing data, models, and scenarios;
- Modeling approaches that are feasible for the 2013 report versus those that become viable in longer-term assessment activities; and
- Ideal modeling approaches versus identifying and using the aspects of an ideal solution that matter the most for addressing particular research and assessment challenges.

For example, one might like to have a fully integrated and validated model that simulates all the anthropogenic and natural forcing of the climate system, including complete descriptions of all feedbacks making decadal and longer simulations of the climate system as credible as understanding permits. But failing that, currently available integrated assessment models are useful tools for integrated analysis of the climate system and the human factors and activities that force climate change. There are still important gaps in basic understanding as well as in applications of what we know. The challenge is to fill these gaps in a sensible and stepwise fashion rather than expecting to find a fully satisfying solution in the short term.

3.2.2 Overview of Modeling Practices in Past U.S. National Assessments

The current National Climate Assessment follows two previous formal national assessments. The first, Climate Change Impacts on the United States (NAST, 2001) attempted to avoid controversies associated with climate scenarios and modeling by using only published results included in assessments by the IPCC. A single emissions scenario (Scenario IS92a of the second IPCC assessment) and results derived using two climate models (the United Kingdom Met Office Hadley Centre and the Canadian Centre for Climate Modelling and Analysis models) were considered along with some exploration of climate model performance, but not an extensive evaluation. There was significant deliberation about the scenarios and models used by the national assessment and the rationale for their selection.

At the time of the first assessment, downscaling was largely accomplished by statistical processing to apply changes in temperature, precipitation, and other variables to the instrumental record of the 20th century. The first national assessment included some ecosystem process modeling and limited dynamic vegetation modeling. It was built around the Vegetation Ecosystem Modeling and Analysis Project (VEMAP), an extensive ecosystem model study. The VEMAP study focused on changes in potential natural vegetation and biogeochemical fluxes, with limited transfer of information to economic models for the agriculture and forestry sectors and limited modeling of other stresses on systems. Other modeling applications included hydrologic models for flow impacts, agricultural productivity modeling, a quantitative exploration of uncertainties, and explicit use of a confidence lexicon based on the expert judgment of the National Assessment Synthesis Team responsible for the assessment.

The second national assessment, entitled *Global Climate Change Impacts in the United States* (Karl et al., 2009) incorporated results summarized by the Synthesis and Assessment Products (SAPs) of the U.S. Climate Change Science Program (CCSP). Unlike the first national assessment, the second assessment did not include significant outreach to stakeholders. The regional chapters summarized and synthesized published information. The confidence lexicon for the second assessment was based on the expert judgment of the authors. The report relied on model results completed for the fourth IPCC assessment. Two different emissions scenarios (high and low), related to those described by the IPCC Special Report on Emissions Scenarios (SRES) (Nakicenovic and Swart, 2000), were considered. The second assessment report described the observational record of climate change in some detail and explained how climate models are evaluated and incorporated results of modeling studies focused on specific sectors, primarily as described by the CCSP Synthesis and Assessment Products.

3.2.3 Challenges, Opportunities and Lessons Learned in Downscaling and Regional Modeling

Because the spatial resolutions of global climate models are too coarse to resolve climate change and impacts at scales relevant to most decisions, stakeholders often require downscaling of results derived from global modeling studies. Although some global models are simulating climate change at finer scales, partially addressing some of these

scale issues, downscaling continues to be required for many applications. Future projections of some variables are relatively insensitive to downscaling method. For others, such as frequency of extreme events, results vary significantly between downscaling methods.

A presentation on statistical and dynamical downscaling provided multiple examples of downscaling efforts that were developed for specific applications, including projections of changes in winter snowfall, extreme heat events, precipitation, heating degree-days, persistence of drought, lake levels, and effects on specific sectors. But the utility of different methods for downscaling climate model results to regional scales within North America needs consistent evaluation. Different downscaling approaches yield significantly different results, but most evaluations of alternative methods have focused only on attributes important to developers rather than those of high priority to users.

Users do not always recognize that particular modeling approaches and methodologies are better suited to certain applications than others. Clearly, some models perform better than others with respect to specific metrics. Reproducing observations does not mean that models represent mechanisms correctly, implying that they may not provide accurate and consistent projections beyond a period of observations. There is a need to know much more about processes that occur on different spatial and temporal scales.

But information incorporated into the NCA, as well as adaptation and mitigation decisions, must be derived using clearly explained and thoroughly evaluated methods. A general approach to evaluating downscaling methods is to
1. Develop a standard set of appropriate physical and statistical tests that can be used to evaluate any statistical or dynamical downscaling method for daily temperature or precipitation;
2. Apply these standard tests to downscaling approaches in current use for impact studies; and
3. Determine whether for a specific location, time scale, and purpose any of these methods can be judged by an objective set of criteria to be better than others.

While some relatively simple downscaling methods are reliable for climatological means, more complex methods are needed to simulate changes in thresholds and extremes. No method, including those involving regional climate models, is guaranteed to correct for global model bias in multi-day events. A few large data sets have been created to facilitate comparisons between models and methods and as a basis for evaluating strengths and weaknesses of different approaches and methods. In addition, combinations of model outputs (ensemble approaches) have been compared to individual model results. Most downscaling approaches, however, yield results better suited to smaller spatial scales than does the direct use of global climate model output. But the best downscaling approach for any given analysis depends on the research question being asked. Thus, understanding the limitations of and biases inherent in different methods is important for selecting a method and determining how to interpret, or whether to use, the results.

A wide variety of stakeholders have asked climate modelers to support downscaling of information for decision support. It was clear to many at the workshop that not all stakeholders understand how best to use such information, and that most modelers find it difficult to identify ways to make their results most useful. Some lessons learned in providing regional-scale climate data in multiple contexts include

- Two-way interactions and long-term relationships with stakeholders are crucial for building trust, understanding the decision context, and defining the right questions to ask and answer;
- Use of state-of-the-art, well-documented climate models and well-evaluated downscaling approaches is important—our understanding of these issues is changing rapidly;
- Both dynamical (process-based) and statistical downscaling approaches have important advantages and limitations;
- There is a need to thoroughly evaluate methods relative to historical observations in order to quantify limitations and robustness at spatial and temporal scales relevant to impacts;
- In order to avoid overemphasizing short-lived trends, it is useful to average results over climatological periods (20–30 years);
- To ensure that users make effective use of climate projections, it is important to cast projections in terms that are relevant to impacts;
- There is a need to address likelihood and ranges of uncertainties in future projections; and
- Communication of findings to the intended audience in their own language is critical.

The North American Reanalysis Climate Change Assessment Program (NARCCAP)[1] represents a significant new source of information for the NCA. Using four global climate models in conjunction with six regional climate models, the NARCCAP program explores multiple uncertainties in regional and global climate model projections. The program is developing multiple high-resolution (50 km) regional climate scenarios for use in assessments of impacts and adaptation. The purpose of NARCCAP is to evaluate regional model performance in order to establish the credibility of individual simulations for the future. There are seven primary institutions participating with funding from a variety of sources.

Phase I of NARCCAP involved testing models against observed information, while Phase II involves testing alternative current and future projections. These efforts allow quantification of uncertainty at regional scales using probabilistic approaches. These comparisons save 53 different variables at three-hour intervals. Three uses are anticipated for this information: (1) further dynamical or statistical downscaling; (2) regional analysis of NARCCAP results; and (3) application of results as scenarios for impacts and adaptation studies. An example of a current application of NARCCAP results is work to develop adaptation plans for Colorado River water resources.

3.2.4 Characterizing Uncertainty

The NCA is treating uncertainty through two alternative paradigms: (1) climate change in which uncertainty is assumed to cascade from global, to regional climate change, and to impact models, or (2) system vulnerability beginning with an uncertainty analysis of a system that is potentially sensitive to climate change through to evaluation of how large climate change has to be to overtake other stresses as a key driver. In either paradigm, it is important to consider:

- Temporal and spatial scales of interest and importance;
- Process understanding;
- Limits on predictability of the real world; and
- Impacts of uncertainty communications on decisions, including lack of a common vocabulary.

The tractability of uncertainty issues depends partly on the time frame for analysis. Near-term predictability in a climate change context is very difficult.

---

[1] http://www.narccap.ucar.edu/

Sensitivity studies can be useful. It is useful to describe uncertainty, not just quantify it. Confusion leads to inaction, and if people think that discussion of uncertainty means that scientists are not certain about what they know, it is likely to have a negative effect on adaptation. The precautionary principle can be used in communications about uncertainty, and climate uncertainty can also be put in the context of other kinds of uncertainty.

The concept of robust decision making (Lempert et al., 2006) de-emphasizes the importance of accurate predictions and focuses on decisions that are worth making in the context of multiple futures. Engineers make decisions under uncertainty every day; there is a need for better understanding of combined sources of uncertainty and how to respond in a rational way.

3.2.5 Relevant Outcomes of the Scenarios Workshop

Key outcomes from the Scenarios Workshop were reported to the group. Among them were that the topic of scenarios itself is extremely complex. There is a lot of confusion about the concept of scenarios and a lot of different perceptions of what a scenario represents. For example, one community's model output can be another community's scenario. The scenarios workshop participants are strongly motivated to do multi-sector and regional analyses that can be related in some way to other global or international analyses, but current capacity is inadequate for such an undertaking.

As with modeling, scenario development for the 2013 report will recognize resource limitations but contribute toward a more ambitious longer-term goal. In addition, there is strong interest in ensuring that scenario activities of the NCA be application-driven, be useful in real world contexts, and be responsive to communications needs. This interest implies the need for a set of alternative approaches based on an inventory and analysis of methods and results that are currently available. There are many scenario-based activities currently in place, but they are extremely diverse and oriented toward a wide range of users, some of whom are end users of information and others who are translators or intermediate users between scientists and decision makers.

3.2.6 Sectoral Modeling Issues

Methodological issues in scaling different families of models were discussed in the context of energy and economic information, land use, water, coastal management, and overall problem solving in moving from one scale to another. Because there are a plethora of different applications with a wide variety of objectives, it is difficult to do intercomparisons. Some intercomparisons have addressed the water sector, with one effort looking at why different models leading to different conclusions about the magnitude of flow reductions in the Colorado River. Another effort identified elasticity of runoff reductions as related to reductions in precipitation. In order to determine socioeconomic and environmental impacts of sea-level rise on the U.S., an integrated coastal modeling effort was suggested.

3.2.7 Scaling Issues

Climate does not change gradually in time or space; there is variability in both dimensions. It is important to recognize that there can be discontinuities. Not all models are created equal. Though a straightforward approach is to average results derived using different models, this is not always the best approach for answering specific questions. There are clearly some models that perform better than others with respect to specific metrics. Reproducing observations does not mean that models represent mechanisms correctly, implying that they may not provide accurate and consistent projections beyond a period of observations. There is a need to know much more about processes that occur on different spatial and temporal scales.

For decision support it is helpful to determine upper and lower bounds of results, but it is important to establish whether that range is consistent with historical data. In many instances, extremes may be underrepresented. Estimates of lower bounds typically are more reliable than those for upper bounds. Unlikely events with high impact must be considered, and it is important to recognize that as climate impacts accelerate, the variance in distributions narrows.

3.2.8 Managing and Meeting User Expectations

It was noted that what is possible to achieve in the short term for the report that is due in 2013 is likely to be limited in scope, but we have an opportunity to make significant progress in the context of building a longer-term approach for the Assessment. It is important to consider what investments in science could improve the longer-term utility of models, as well as understanding how what we know now can best be used to increase resilience to climate variability and change. There is a need to advance the ability to use models for evaluating our understanding as the science evolves at the same time that we advance the capacity to appropriately use models for informing a variety of decision-making processes.

3.3 Discussion Group Reports

In order to facilitate focused discussions, workshop participants divided into smaller groups. Six discussion groups convened, three during each of two sessions. Specific questions guided the deliberations of each group. The following sections list these questions and summarize key discussion points and recommendations.

3.3.1 Discussion Group 1. Implications of Scenario Outcomes for Climate Modeling

3.3.1.1 Questions
What do we understand about the desires and needs of various user communities for climate modeling results and their desires and needs for climate modeling results?
- What data from climate models will be available to the NCA?
- How will the Assessment address the desire to use emissions scenarios that differ from those with a climate modeling perspective?
- How might the Assessment use past results (e.g., from the fourth IPCC assessment) versus those developed by Phase 5 of the Coupled Model Intercomparison Project (CMIP5) or the fifth IPCC assessment?
- What are the tradeoffs between using fewer models for consistency versus a selection of models that may better match specific user needs for outputs or spatial and temporal resolutions? What criteria can be used to evaluate such tradeoffs?
- How might model uncertainties be analyzed and included in the Assessment in consistent ways?
- Where can we expect the big gaps in climate modeling? What additional global modeling research is needed to meet long-term NCA needs?

3.3.1.2 Discussion points and suggestions
- As a practical matter, the 2013 report should focus on the results of CMIP Phase 3. The 2013 report might include a chapter contrasting CMIP Phase 3 and CMIP Phase 5. This recommenda-

tion responds, in part, to the realities of limited access to new data within the time frame for the development of the 2013 report, lack of literature describing CMIP5 and its results, and the need to engage a broad group of data users. Unique results expected of CMIP5 include insights about hurricanes and other phenomena needing high resolution analyses as well as about impacts on natural systems such as the global carbon cycle, land-use and land-cover change, and vegetation attributes and processes.
- An evaluation as to which models are reliable at regional scales is needed in order to inform longer-term analyses as well as regional users.
- Increased focus on attribution and understanding climate mechanisms will help determine model validity.
- There is a role for expert elicitation with regard to penetrating the complex landscape of model and scenario alternatives and outputs in order to develop a collective view of alternative climate futures called climate narratives.
- An archive of global model results suitable as input to regional models may be a useful short-term product of the NCA.
- Assess whether excessive focus on regions might miss larger-scale gradients.
- Consider issuing a best practices document for the NCA.
- Regional or sectoral assessments may need a guidance document to ensure they at least initiate similar conversations. Otherwise, there may be no comparability across sectors and regions.
- Consider whether pattern scaling is a useful approach for examining climate change on adaptation time scales? Pattern scaling may help address questions about which scenarios to use.
- The uncertainty in scenarios (e.g., that associated with human activities) may dominate other sources of error.
- The probability distribution function of climate sensitivity has a fat tail not captured by models currently in use. Climate model studies are not capturing the full range of future extremes, especially on the high end.
- It is important not to combine socioeconomic and scientific uncertainty into a single category.
- In a decision context, consider the benefits of characterizing versus quantifying uncertainty. The NCA should consider various approaches to characterizing uncertainty.
- Those involved in the NCA need guidance about access to modeling data archives by experts capable of translating technical descrip-

tions and metadata into useful terms for the Assessment.

- On the longer time frame, future evaluations could focus on important gaps like the future of the Greenland Ice Sheet, higher resolution in space and time, etc.

- The NCA should characterize different types of users of Assessment products.

3.3.2 Discussion Group 2. Implications of Scenarios and Outcomes for Integrated Assessment Modeling and Interactions with Adaptation or Mitigation Modeling

3.3.2.1 Questions

- What do we currently understand about various user communities and their desires or needs for climate modeling results? What data from models will actually be available? How might the Assessment use past results (e.g., Clarke et al., 2007) versus more current results?

- How will the Assessment contend with the use of emissions scenarios that differ from those used in climate modeling studies?

- How close are we to having integrated assessment models with resolutions on geographic scales of particular interest to the NCA? Is statistical downscaling of integrated assessment model results feasible?

- Are there timing issues relevant to the production of model output?

- For impact, adaptation, or mitigation modeling and the use of other models, how consistent does the linking of other models with climate and integrated assessment modeling have to be?

- How will uncertainties be analyzed and model results evaluated in consistent ways?

- Where are big gaps (uncertainties too high or process outputs not available) expected? What additional modeling research is needed to meet long-term NCA needs?

3.3.2.2 Suggestions

- The integrated assessment modeling community is more familiar with certain users than with others. For example, integrated assessment modelers have been working for decades with the energy sector. The water management community is very sophisticated in dealing with climate issues while others are not as engaged.

- There are a lot of communications issues with regard to models and their application. Building relationships and common understanding among and across modelers and user communi-

ties will be challenging. It is unclear whether stakeholders and end users are fully capable of framing their questions so as to be effective in communicating with modeling communities. There is a need to manage expectations.

- There is a need to work with end users to
  ○ Identify variables that can and should be treated within models;
  ○ Determine how to incorporate multiple stressors; and
  ○Frame things such that both adaptation and mitigation questions and decisions are considered.

- Users want to know the scope and magnitude of climate forcing. How disruptive will climate change be? Is change irreversible? Will cumulative impacts affect value streams? Is gradual adaptation possible or will thresholds make this less useful?

- One approach to complexity in scenarios is to employ expert elicitation. Some users would like to see more expert judgment used, even if it is not entirely correct in the end.

- Each project has different capacities, issues, actors and local researchers involved. There is a proliferation of approaches for different situations.

- There are a few good analytical threads that could be used to illustrate some important issues for the NCA. These threads will address primarily the intersection of mitigation challenges and impacts that depend on the allocation of land and water. Assessment of these topics needs to take advantage of the work and the data that the CMIP5 and IPCC Representative Concentration Pathway (RCP) processes and their heritages have already generated.

- Changing land use has feedbacks on crop yields, crop prices, land prices, trade and other macroeconomic consequences. Depending on policy assumptions, the consequences shown by integrated assessment models vary substantially. For the NCA, policy frames are important because they drive these decisions, e.g., the U.S. corn ethanol policy.

- Useful paradigms for the Assessment include statistically downscaled representations of land-use pressures under different mitigation options and regional integrated models. Work to apply the GCAM integrated assessment model to Great Lakes and Gulf Coast regions is already under way.

- The capacity for new modeling studies in support of the Assessment is probably limited at best.
- An inventory of scenarios and underlying data availability is needed.
- For mitigation scenarios, all the emissions, energy technology, concentration data underlying the Representative Concentration Pathways for CMIP5 are available through a Web interface maintained by the International Institute for Applied Systems Analysis or from the modeling groups that developed the RCPs. These data include downscaled land-use data, calculated demands for agricultural products, forest products, and energy and could be adequate to drive other impact models and has been used in this manner.
- Multiple factors influence the decision to include a policy case within the NCA, including the issue that CMIP3 data do not support this. It may be feasible to use past results, e.g., CCSP Synthesis and Assessment Product 2.1 (Clarke et al., 2007).
- In the short term, there are multiple new data sources available to the Assessment including global tree cover (Hansen et al., 2003), more highly-resolved spatial data from other communities (e.g., the Soil Survey Geographic (SSURGO) soils database), USGS land use projections, and U.S. National Park Service socioeconomic narratives.
- Work is underway to
  - Better understand shorter time-scale information;
  - Incorporate climate feedbacks into integrated assessment models,
  - Closely couple hydrology, crop, ecosystem, and climate models; and
  - Develop and apply state-level models, e.g. for California.

But the nature and timing of these development activities makes the delivery of fully-integrated information very challenging in the near term.

- Evaluation of uncertainties in all model applications is important. There is a rich history of impact model evaluation and intercomparisons.
- Research is needed on geographically disaggregated, spatially-specific interactions within both the climate and impact modeling communities.
- Clearly there are some products available now and a few analyses from different groups that might be drawn upon to illustrate interactions between sectors: energy, water, and land;

biofuels and crops; carbon sequestration and ozone interactions; etc.

- There are numerous opportunities to use outputs from the integrated assessment models (e.g., demand for agricultural production, land use allocation, or biofuels) as input to analyses of other sectors such as agriculture, forests, ecosystems, and possibly hydrology, but consistency will be a challenge.
- In some cases, the NCA should consider producing frameworks, such as best practices, that can be used by a number of different communities.

3.3.3 Discussion Group 3. Implications of Scenarios and Outcomes for Regional Climate Modeling and Interactions with Sectors

3.3.3.1 Questions

- What are the particular classes of user (assessment) needs, and what features and characteristics depend on regional modeling outputs?
- What types of regional models are particularly well suited for various classes of user needs based on model outputs and ability to handle various spatial and temporal scales? Are there recognized differences in models within these model types for addressing particular classes of user needs (e.g., temporal scales in general and sector parameters of interest or regional spatial scales)? Does the community have ways of evaluating the effectiveness of each for different types of applications?
- Can a standardized process be developed for analyzing the capabilities and limitations of regional models for assessment purposes (capturing variability and extremes not just the means)?
- What tradeoffs are involved in using fewer models for consistency (for example, larger spatial runs covering more than one region) versus models that may track better with specific user needs for outputs at spatial or temporal scales and topography important for the region? What are the criteria that could be used to make such tradeoff choices?
- How will uncertainties in the models be analyzed and included in the Assessment in consistent ways?
- Where can we expect the big gaps (uncertainties too high or process outputs not available)? What additional research for the global models is needed to meet long-term NCA needs?

3.3.3.2 Discussion points and suggestions

- Guidelines for maintaining consistency between regional and sectoral components of the Assessment are needed. Examples of poor practice are useful.
- Clarify which decisions require downscaled information and which do not.
- Develop strong interactions between those using models to analyze impacts and users of the results.
- Evaluation is related to the question you are asking and the degree of agreement within the community about best practices. Could we provide alternative metrics for consideration?
- A human interface to assessment information and results is needed. Published information cannot anticipate all issues and questions that arise in the use of assessment results.
- Inventories that support coordination within regions are needed. Support to connect users with data sets or assessment activities is needed.
- Climate outlooks could be decision-focused outlooks that provide some of the basis for a broader activity in the future.
- Types of models available to the NCA need to be clarified, and these types need to be related to different types of users.
- An approach for evaluating model capabilities for simulating unexpected or unknown surprises needs to be identified.
- Models should be evaluated with respect to their intended use. It is important that models not be forced into applications that they were not designed or intended for.
- Prepare tutorials for existing resources, but recognize that existing resources and associated documents are often overlooked.
- The NCA needs to provide consistent population projections and needs to develop capabilities for characterizing socioeconomic attributes and variables within regions.
- A climate outlook approach that includes expert integration or elicitation can help incorporate emerging knowledge, such as results from CMIP5.

3.3.4 Discussion Group 4. Tracking Uncertainties in the Assessment System in the Short Term

3.3.4.1 Discussion points and suggestions

- There are at least two paradigms for uncertainty with respect to climate change impacts: (1) a climate change paradigm, and (2) a system vulnerability paradigm. These two paradigms are not mutually exclusive, but they can differ in very important ways.
- The climate change paradigm implies a cascade of uncertainty from global climate model results through regional climate and sectoral models to results.
- Under the system vulnerability paradigm, sectoral decisions are made in the context of a suite of stresses, of which climate is just one. While climate is important, uncertainties must be evaluated for all stresses. In a context of multiple stresses, uncertainties with respect to climate are considered in the context of uncertainties with regard to society, government, other aspects of physical systems, human health, and other multiple stresses.
- It is useful to consider for a given system, how large climate change would have to be in order to dominate other stressors.
- It is useful to describe, not just quantify, uncertainty. A rigorous uncertainty vocabulary is important.
- In either paradigm, it is important to consider
  - Temporal and spatial scales;
  - Process understanding;
  - Limits of real world predictability; and
  - Diversity within the audience for assessment results.

3.3.5 Discussion Group 5. Matching Model Outputs or Performance to Needs from User Communities and Decision Makers

3.3.5.1 Questions

- What factors are important to matching model outputs or performance to the needs of user communities and decision makers?
- How do we meet the challenges of identifying different decision makers and stakeholders and of structuring a process for communicating their needs effectively to the appropriate modeling communities, both climate and otherwise?
- How do we manage expectations to what is actually scientifically feasible and show how both communities can learn by this?

3.3.5.2 Discussion points and suggestions

- There is concern about focusing a research program on narrowing uncertainties, as opposed to a path forward to use what we know now.
- Understanding historical background is part of being prepared for extreme events—this topic

evolved from a discussion about understanding natural system thresholds or tipping points.

- We have an infrastructure for engagement with users now that did not exist before, e.g., the Regional Integrated Science and Assessment Program. A useful approach is to start with important vulnerabilities identified by previous assessments and then provide a suite of general questions associated with those vulnerabilities that can be addressed as high priorities.

- With respect to integrated assessment approaches, if we do not consider feedbacks, we will be overly optimistic about outcomes. For example, the agricultural economy is global, and if we consider only climate inputs to U.S. agriculture, we will not account for exogenous drivers.

- Storylines are useful for decision makers. Decision makers can use storylines for strategic planning without a lot of quantification—this approach has less potential for misuse of information or misunderstanding how to use information.

- The implications of NCA support for national policy development need to be clarified.

- Both the assessment and research communities evolve continuously, and it is important to recognize that there will not be a single answer for any given question.

3.3.6 Discussion Group 6. Transitioning from the Present to Future Capabilities

3.3.6.1 Questions
- What should the future look like?
- How can we transition into the future?

3.3.6.2 Discussion points and suggestions
- Future capabilities should include a system of downscaled climate data. The interface between global climate models and regional models needs to be more explicit and rigorous. Formal evaluations of climate models, downscaling methods, and tools for evaluating impacts are needed.

- The use of observations needs to be strengthened, including the development of standard indicators and measures of model accuracy and veracity. Guidance by the climate modeling community regarding the best use of climate data to address specific questions is needed.

- The transition to a more effective National Climate Assessment process requires planning for a stronger architecture. Agencies need to give higher priority to the NCA and provide more

resources. A National Center for Climate Assessment with public engagement, cross-laboratory capacity, cross-discipline intellectual resources, and cross-community capacity may be a useful entity for assembling the required resources and personnel.

3.4 Next Steps and Potential Actions

A summary of key suggestions from workshop participants follows.

3.4.1 Decision Support
- Generate regional climate outlooks that help people understand the alternative climate futures they may need to prepare for. These regional outlooks can be composites of information derived from a variety of models and provided by modeling experts combined with expert elicitation by regional experts.

- Include explorations of different elements of vulnerability beyond just climate change.

- Complete an inventory of the downscaling efforts that are currently underway and a rigorous intercomparison and evaluation of downscaling methods. One possible approach is a National Academy of Sciences study, but the Assessment can probably make contributions in the best practices category.

- Create a catalog of principles for evaluating impact, adaptation, and vulnerability studies for regions and sectors of interest.

- Take advantage of the joint Adaptation Task Force of the Council on Environmental Quality, the Office of Science and Technology Policy, and the National Oceanic and Atmospheric Administration.

- Hold a semi-annual conference to provide a venue for presenting and sharing experiences in the use of model results.

- Develop more capacity to provide model information at decision scales and help users with science translation and appropriate use of modeling information, including asking the right questions.

3.4.2 Science Support Context
- Use the CMIP3 archive of model solutions as primary climate projection data for the 2013 NCA, but to the extent feasible include comparisons with the outcomes from the most recent modeling efforts such as CMIP5, at least some kind of case study, as part of the NCA efforts as well. This approach recognizes both

the quantity and quality of work based on the older models and the fact that the timing of this Assessment is not ideal for deployment of new model results.

- Develop a program that focuses on model evaluation and attribution (explanation of climate mechanisms) of trends in climate as well as extreme events.
- Help support the development of a boundary organization or network of such organizations to link users to fundamental research communities in more effective ways, including identifying important research needs, for example, a group to provide feedback between ecologists and climate modelers regarding the ability of the models to represent persistent drought.

3.4.3 Contending with Uncertainty

- Over the short term, there is a need to manage expectations about our ability to generate consistent model output from a range of climate, socioeconomic, sectoral and integrated models. Effective evaluation of uncertainty is a long-term objective that is very difficult to accomplish at this point.
- There is a need for transparent reporting of sources of uncertainty throughout the Assessment process, from measurement error to structural uncertainty in models, etc.
- The Assessment document needs to discuss climate model evaluation at various scales, including model strengths and weaknesses. It should also have sections that evaluate modeling approaches and capacity for impact sector models and integrated models.
- Uncertainties need to be discussed, but it is also useful and empowering to focus on how we can move ahead based on what we do know.
- Scale is important for a variety of reasons, including the point that as scale increases, new processes dominate, e.g., in the ecosystem analogy, individual trees are modeled at the smallest scales.
- The NCA should begin an evaluation of which uncertainties matter in particular decision-making contexts as well as in our scientific understanding of the components of the Earth system.
- The Assessment should be clear about what scientific credibility really means—it is not just about published results.

3.4.4 Next Steps

- Treat activities and tasks required to prepare the 2013 Assessment as stepping-stones towards a longer-term Assessment process.
- Plan a participatory process for scenario creation that includes climate and socioeconomic components that can be understood in broader national and international contexts, for example useful for IPCC assessments or suitable for nesting within other larger-scale assessments.
- For the 2013 report, develop a broad national framework for model information that can support regional details of a relatively simple nature.
- Incorporate a policy case into scenarios that includes greenhouse gas management, e.g., by adopting different stabilization scenarios from the IPCC Representative Concentration Pathways or examples from the CCSP Synthesis and Assessment Product 2.1 (Clarke et al., 2007). Adding a policy case will allow consideration of interactions between mitigation and adaptation efforts.
- Limit the number of inter-sectoral studies to topics for which sufficient capacity exists. Suggestions include
  - Water, energy, and land in the context of biofuels;
  - Carbon storage, ecosystem processes, and air pollution; and
  - Sea-level rise, river outflow, and hypoxia.
- Consider unanticipated outcomes that are the result of inter-sectoral interactions, e.g., competition for land and water in changing socioeconomic conditions.
- Increase collaboration between government agencies to maximize the effectiveness of scarce funding and to demonstrate needs for larger budgets in the future.
- Take maximum advantage of research grant opportunities to address key NCA research questions.

REFERENCES

Ammann, C. M., F. Joos, D. S. Schimel, B. L. Otto-Bliesner, and R. A. Tomas (2007), Solar influence on climate during the past millennium: Results from transient simulations with the NCAR Climate System Model, *Proceedings of the National Academy of Sciences of the United States of America*, 104(10), 3713-3718.

CCSP (2008), [1]*The Effects of Climate Change on Agriculture, Land Resources, Water Resources, and Biodiversity in the United States*, A Report by the U.S. Climate Change Science Program and the Subcommittee on Global Change Research. P. Backlund, A. Janetos, D. Schimel, J. Hatfield, K. Boote, P. Fay, L. Hahn, C. Izaurralde, B. A. Kimball, T. Mader, J. Morgan, D. Ort, W. Polley, A. Thomson, D. Wolfe, M. G. Ryan, S. R. Archer, R. Birdsey, C. Dahm, L. Heath, J. Hicke, D. Hollinger, T. Huxman, G. Okin, R. Oren, J. Randerson, W. Schlesinger, D. Lettenmaier, D. Major, L. Poff, S. Running, L. Hansen, D. Inouye, B. P. Kelly, L. Meyerson, B. Peterson, and R. Shaw. United States Department of Agriculture, Washington, D.C., USA, 362 pp.

Clarke, L., J. Edmonds, H. Jacoby, H. Pitcher, J. Reilly, and R. Richels (2007), *Scenarios of Greenhouse Gas Emissions and Atmospheric Concentrations*. Sub-report 2.1A of Synthesis and Assessment Product 2.1 by the U.S. Climate Change Science Program and the Subcommittee on Global Change Research. U.S. Department of Energy, Office of Biological and Environmental Research, Washington, D.C., USA, 106 pp.

Hansen, M. C., R. S. DeFries, J. R. G. Townshend, M. Carroll, C. Dimiceli, and R. A. Sohlberg (2003), Global percent tree cover at a spatial resolution of 500 meters: First results of the MODIS vegetation continuous fields algorithm, *Earth Interactions*, 7(10), 1-15.

Hawkins, E., and R. Sutton (2009), The potential to narrow uncertainty in regional climate predictions, *Bulletin of the American Meteorological Society*, 90(8), 1095-1115.

Hawkins, E., and R. Sutton (2010), The potential to narrow uncertainty in projections of regional precipitation change, *Climate Dynamics*, 37(1-2), 407-418.

Hayhoe, K. (2010), "A Standardized Framework for Evaluating the Skill of Regional Climate Downscaling Techniques" (Ph.D. dissertation, University of Illinois at Urbana-Champaign).

IPCC (2001), *Climate Change 2001: Mitigation : Contribution of Working Group III to the Third Assessment Report of the Intergovernmental Panel on Climate Change*, Mertz, B., O. Davidson, R. Swart, and J. Pan (Eds.), Cambridge University Press, Cambridge, U.K., 753 pp.

Janetos, A. C., L. Clarke, W. Collins, K. Ebi, J. Edmonds, I. Foster, H. J. Jacoby, K. Judd, L. Leung, R. Newell, D. Ojima, G. Pugh, A. Sanstad, P. Schultz, R. Stevens, J. Weyant, T. Wilbanks, M. Knotek, and E. Malone (2009), *Science Challenges and Future Directions: Climate Change Integrated Assessment Research*. U.S. Department of Energy, Office of Biological and Environmental Research, Washington, D.C., USA, 80 pp.

Karl, T. R., J. M. Melillo, and T. C. Peterson (Eds.) (2009), *Global Climate Change Impacts in the United States*, Cambridge University Press, Cambridge, U.K., 188 pp.

Lempert, R. J., D. G. Groves, S. W. Popper, and S. C. Bankes (2006), A general, analytic method for generating robust strategies and narrative scenarios, *Management Science*, 52(4), 514-528.

Liang, X.-Z., K. E. Kunkel, G. A. Meehl, R. G. Jones, and J. X. L. Wang (2008), Regional climate models downscaling analysis of general cirulation models present climate biases propagation into future change projections, *Geophysical Research Letters*, 35, L08709, 1-5.

Melillo, J. M., J. Borchers, J. Charney, H. Fisher, S. Fox, A. Haxeltine, A. Janetos, D. W. Kicklighter, T. G. F. Kittel, A. D. McGuire, R. McKeown, R. Neilson, R. Nemani, D. S. Ojima, T. Painter, Y. Pan, W. J. Parton, L. Pierce, L. Pitelka, C. Prentice, B. Rizzo, N. A. Rosenbloom, S. Running, D. S. Schimel, S. Sitch, T. Smith, and I. Woodward (1995), Vegetation ecosystem modeling and analysis project— Comparing biogeography and biogeochemistry models in a continental-scale study of terrestrial ecosystem responses to climate-change and CO_2 doubling, *Global Biogeochemical Cycles*, 9(4), 407-437.

Morgan, M. G., R. Cantor, W. C. Clark, H. D. Fisher, H. D. Jacoby, A. C. Janetos, A. P. Kinzig, J. Melillo, R. B. Street, and T. J. Wilbanks (2005), Learning from the U.S. national assessment of climate change impacts, *Enviornmental Science and Technology*, 39(23), 9023-9032.

Moss, R. H., and S. H. Schneider (2000), Uncertainties, in *Guidance Papers on the Cross Cutting Issues of the Third Assessment Report of the IPCC*, edited by R. Pachauri, T. Taniguchi and K. Tanaka, World Meteorological Organization, Geneva, 33-51.

Moss, R. H., J. A. Edmonds, K. A. Hibbard, M. R. Manning, S. K. Rose, D. P. van Vuuren, T. R. Carter, S. Emori, M. Kainuma, T. Kram, G. A. Meehl, J. F. B. Mitchell, N. Nakicenovic, K. Riahi, S. J. Smith, R. J. Stouffer, A. M. Thomson, J. P. Weyant, and T. J. Wilbanks (2010), The next generation of scenarios for climate change research and assessment, *Nature*, 463(7282), 747-756.

Nakicenovic, N., and R. Swart (Eds.) (2000), *Special Report on Emissions Scenarios: a special report of Working Group III of the Intergovernmental Panel on Climate Change*, Cambridge University Press, New York, NY, USA.

NAST (2001), *Climate Change Impacts on the United States: The Potential Consequences of Climate Variability and Change*, National Assessment Synthesis Team, Cambridge University Press, Cambridge, U.K.

Parson, E. A., R. W. Corell, E. J. Barron, V. Burkett, A. Janetos, L. Joyce, T. R. Karl, M. C. MacCracken, J. Melillo, M. G. Morgan, D. S. Schimel, and T. Wilbanks (2003), Understanding climatic impacts, vulnerabillities, and adaptation in the United States: building a capacity for assessment, *Climatic Change*, 57(1-2), 9-42.

Reilly, J., and S. Paltsev (2009), Biomass energy and competition for land, in *Economic Analysis of Land Use in Global Climate Change Policy*, edited by T. Hertel, S. Rose and R. Tol, Routledge, Oxford, U.K., 184-207.

Savonis, M. J., V. R. Burkett, and J. R. Potter (Eds.) (2008), *Impacts of Climate Change and Variability on Transportation Systems and Infrastructure: Gulf Coast Study, Phase I*, United States Department of Transportation, Washington, D.C., USA.

Scott, M. J., R. D. Sands, J. Edmonds, A. M. Liebetrau, and D. W. Engel (1999), Uncertainty in integrated assessment models: Modeling with MiniCAM 1.0, *Energy Policy*, 27(14), 855-879.

Taylor, K. E. (2001), Summarizing multiple aspects of model performance in a single diagram, *Journal of Geophysical Research*, 106(D7), 7283-7192.

Thomson, A. M., K. V. Calvin, L. P. Chini, G. Hurtt, J. A. Edmonds, B. Bond-Lamberty, S. Frolking, M. A. Wise, and A. C. Janetos (2010), Climate mitigation and the future of tropical landscapes, *Proceedings of the National Academy of Sciences of the United States of America*, 107(46), 19633-19638.

Thomson, A. M., K. V. Calvin, S. J. Smith, G. P. Kyle, A. Volke, P. Patel, S. Delgado-Arias, B. Bond-Lamberty, M. A. Wise, L. E. Clarke, and J. A. Edmonds (2011), RCP4.5: a pathway for stabilization of radiative forcing by 2100, *Climatic Change*, 109(1), 77-94.

Vrac, M., M. Stein, K. Hayhoe, and X. Liang (2007), A general method for validating statistical downscaling ethods under future climate change, *Geophysical Research Letters*, 34, L18701, 1-5.

Wise, M., K. V. Calvin, A. M. Thomson, L. E. Clarke, B. Bond-Lamberty, R. D. Sands, S. J. Smith, A. C. Janetos, and J. A. Edmonds (2009), The implications of limiting CO_2 concentrations for land use and energy, *Science*, 324(5931), 1183-1186.

**Climate Change Modeling and Downscaling Workshop -
Issues and Methodological Perspectives for the
U.S. National Climate Assessment**

Goal of the Workshop: *The principal goal of this workshop is to provide a current status of science, tools, and capabilities for developing, using, and evaluating climate, integrated assessment, and other models as applicable for the U.S. National Climate Assessment. This will include a specific targeted discussion of issues associated with regional downscaling of various modeling results. The report will provide foundational insights that can help shape subsequent recommendations and guidance to analysis teams engaged in the U.S. National Climate Assessment.*

Dates: December 8-10, 2010
Duration: 2 ½ days, first half-day follows the Scenarios Workshop
Location: Hyatt Arlington, in Rosslyn (Arlington, VA)

Day 1, Wednesday, December 8th: Scenarios, Modeling, and User Driven Needs and Scales

1:00 p.m.: Welcome and Introduction: Tony Janetos

1:10 p.m.: Workshop Charge and Process Overview: Bob Vallario

1:20 p.m.: Goals for the Meeting, Review of Agenda, Expected Products and Outcomes, Overall Timeline: Tony Janetos

1:30 p.m.: National Climate Assessment: Goals, Process, Schedule and Implications for this Workshop: Kathy Jacobs

2:00 p.m.: An Overview of Past Practices and Challenges for the Present for the use of models in US National Assessments: Tony Janetos

- Climate models, IAMs, Impact Sector models
- Importance of different models and spatial scales of analysis for different kinds of decisions
- Importance of considering different temporal scales of analysis
- Characterizing Uncertainties
- Distinguishing the use of scenarios from technical modeling issues

2:30 p.m.: An Overview of Past Practices and Challenges for the Present for scaling the output of climate models in Assessments: Katharine Hayhoe

- Statistical and Dynamic Downscaling
- Regional Modeling
- Fine-scaled global modeling
- Shorter vs. longer time-scales (i.e. century long runs vs. runs of a few years)

3:00 p.m.: Outcomes from Scenarios Workshop Relevant to Modeling Workshop: Richard Moss

- Scaling and characteristics of user needs
- Narratives in the context of scenarios
- Participatory processes
- Characterizing uncertainty in scenarios

3:30 p.m.: Break

3:45 p.m.: Panel: Issues in Downscaling Climate Information - What's Been Done Before and Lessons
Learned for future activities: Bill Collins, Chair
Each presentation will be no more than 10 minutes long, and will possibly allow one clarifying question.
Group discussion follows the last presentation, and can proceed for 45 minutes.

* Examples from Previous National and Regional Assessments: Katharine Hayhoe
* Experience from NARCCAP: Ruby Leung
* Matching climate information with user needs in Chicago: Don Wuebbles
* Assessing and characterizing uncertainties: Noah Diffenbaugh
* Concepts for future development in climate downscaling: Ken Kunkel

5:30 p.m.: Discussion of Afternoon Session (in plenary)

6:15 p.m.: Adjourn

Day 2, Thursday, December 9th: Scales and Modeling Issues

Morning: Opening Session

8:30 a.m.: Summary of Previous Day's Discussion and Major Points and Introduction to Today's Agenda:
Tony Janetos

8:45 a.m.: Sectoral and regional needs for modeling in the NCA – short term and longer: Fred Lipschultz

9:05 a.m.: Discussion of Implications for NCA Modeling: Kathy Hibbard, Moderator

9:35 a.m.: Charge to Breakout Groups: Three breakout groups to discuss implications of scenario outcomes
for climate modeling and for IA/Sectoral/Regional modeling; breakout groups will be charged with reporting
back on their recommendations for what could be done in the short-term for the first NCA, and in the
longer-term, with a specific focus on what types of research investments could be made by the agencies:
Katharine Hayhoe to deliver charge

* Implications for Climate Modeling in NCA: Bill Collins, Chair
* Implications for IA and interactions with adaptation/mitigation modeling in NCA: Tony Janetos,
  Chair
* Implications for Regional Climate Modeling and interactions with Sectors: Don Wuebbles, Chair

12:00 p.m.: Lunch

1:30 p.m.: Report back from Breakout Groups (15 minutes each from chairs as a panel, followed by 30
minutes for Q&A

2:45 p.m.: Break

3:15 p.m.: Panel: Methodological Issues in Scaling Different Families of Models (15 minute panel
presentations and needs for appropriate input information plus 30 minutes for discussion) – Bill Collins,
Chair

* Matching scales on energy/economic information: Leon Clarke
* Matching scales on land-use information: the RCP Perspective: George Hurtt
* Matching scales on water sector information: Dennis Lettenmeier
* Matching scales for a coastal perspective: Paul Kirshen
* Matching scales from a problem-solving perspective: Keith Dixon

5:15 p.m.: Summary and Tasks for Last Day: Tony Janetos

5:30 p.m.: Adjourn

Evening: Organizing Committee Meets and Starts Work on Synthesis Presentation

Day 3, Friday, December 10th: Current Capacities, Vexing Issues and What is Needed for the Longer Term

8:30 a.m.: Overview of the Topics of the Day: Bill Collins

- Current capacities of AOGCMs and ESMs for use in NCA
- Interface with AR5, CMIP5 – Status of the RCP Process and Simulations
- Towards multi-sector, regional integrated modeling
- Characterization of uncertainties

9:00 a.m.: AOGCM's/ESMs and Interface with RCPs in AR5 and CMIP5: Current status, expected outputs and schedule: Dave Bader

9:30 a.m.: IAMs, their role in AR5 and their possible use in NCA: Jae Edmonds

10:00 a.m.: Sector Models 1: Water: Dennis Lettenmeier

10:30 a.m.: Sector Models 2: Coastal: Paul Kirshen

11:00 a.m.: Assessing Model Uncertainties: Bill Collins

11:30 a.m.: Managing User Expectations in the Light of Scientific Progress: Kathy Jacobs

12:00 p.m.: Charge to Breakout Groups: Tony Janetos

- Tracking uncertainties in the assessment system in the short term - (Noah Diffenbaugh, Chair)
- Matching model outputs/performance to needs from user communities and decision makers - (Kathy Hibbard, Chair)
- Transitioning from the present to future capabilities - (George Hurtt, Chair)

Groups meet from 12:00-2:30 p.m.; including time for lunch

2:45 p.m.: Afternoon Plenary: Reports back from Breakout Groups and Discussion (10 minute reports from each group plus 30 minutes for discussion)

4:00 p.m.: Synthesis Presentation from Organizing Committee: Tony Janetos and Discussion

5:00 p.m.: Adjourn

APPENDIX B. PARTICIPANT LISTS

B.1 Organizing Committee

Tony Janetos, Joint Global Change Research Institute
Don Wuebbles, University of Illinois
Bill Collins, Lawrence Berkeley National Laboratory
Noah Diffenbaugh, Stanford University
Katharine Hayhoe, Texas Tech University
Kathy Hibbard, Pacific Northwest National Laboratory
George Hurtt, University of Maryland

B.2 Workshop Participants

Don Anderson, National Oceanic and Atmospheric Administration
Jeff Arnold, U.S. Army Corps of Engineers
Raymond Arritt, Iowa State University
Dave Bader, Oak Ridge National Laboratory
Venkatramani Balaji, National Oceanic and Atmospheric Administration
John Balbus, National Institutes of Health
Anjuli Bamzai, National Science Foundation
Dan Barrie, University of Maryland
David Behar, San Francisco Public Utilities Commission
Levi Brekke, U.S. Bureau of Reclamation
Jim Buizer, Arizona State University
Lawrence Buja, National Center for Atmospheric Research
Dan Cayan, Scripps Institution of Oceanography
Leon Clarke, Pacific Northwest National Laboratory
Bill Collins, Lawrence Berkeley National Laboratory
David Considine, National Aeronautics and Space Administration
Ben DeAngelo, U.S. Environmental Protection Agency
Eric DeWeaver, National Science Foundation
Noah Diffenbaugh, Stanford University
Keith Dixon, National Oceanic and Atmospheric Administration
James Edmonds, Pacific Northwest National Laboratory
Ron Ferek, U.S. Navy
Gerald Geernaert, U.S. Department of Energy
Aris Georgakakos, Georgia Tech University
Bryce Golden-Chen, U.S. Global Change Research Program
Anne Grambsch, U.S. Environmental Protection Agency
John Hall, U.S. Department of Defense
Murali Haran, Penn State University
Katharine Hayhoe, Texas Tech University
Isaac Held, National Oceanic and Atmospheric Administration
Kathy Hibbard, Pacific Northwest National Laboratory
Justin Hnilo, National Oceanic and Atmospheric Administration
George Hurtt, University of Maryland
Kathy Jacobs, Office of Science and Technology Policy
Tony Janetos, Pacific Northwest National Laboratory
Renu Joseph, U.S. Department of Energy
Thomas Karl, National Oceanic and Atmospheric Administration
Paul Kirshen, Battelle Memorial Institute

Dorothy Koch, U.S. Department of Energy

Kenneth Kunkel, National Oceanic and Atmospheric Administration

Dennis Lettenmeier, University of Washington

L. Ruby Leung, Pacific Northwest National Laboratory

Maxine Levin, U.S. Department of Agriculture

Xin-Zhong Liang, University of Illinois

Fred Lipschultz, U.S. Global Change Research Program

Mike MacCracken, Climate Institute

Julie Maldonado, U.S. Global Change Research Program

Ed Maurer, Santa Clara University

Linda Mearns, National Center for Atmospheric Research

Jerry Melillo, Marine Biological Laboratory

Richard Moss, Pacific Northwest National Laboratory

Philip Mote, Oregon Climate Change Research Institute and Oregon Climate Services

Ramakrishna Nemani, NASA Ames Research Center

Sheila O'Brien, U.S. Global Change Research Program

Robin O'Malley, U.S. Geological Survey

Dennis Ojima, Colorado State University

Zaitao Pan, St. Louis University

Sara Pryor, Indiana University

Richard Rood, University of Michigan

Cynthia Rosenzweig, NASA Goddard Institute for Space Studies

Edmond Russo, U.S. Army Corps of Engineers

Glenn Rutledge, National Oceanic and Atmospheric Administration NCDC

Marcus Sarofim, U.S. Environmental Protection Agency

Elena Shevliakova, Princeton University

Benjamin Sleeter, U.S. Geological Survey

Ronald Stouffer, National Oceanic and Atmospheric Administration

Max Suarez, National Aeronautics and Space Administration

Adam Terando, North Carolina State University

Bob Vallario, U.S. Department of Energy

Cameron Wake, University of New Hampshire

Dan Walker, Computer Science, Corp.

Anne Waple, National Oceanic and Atmospheric Administration

Michael Wehner, Lawrence Berkeley National Laboratory

Tom Wilbanks, Oak Ridge National Laboratory

Don Wuebbles, University of Illinois

Kandis Wyatt, National Oceanic and Atmospheric Administration

Zhaoqing Yang, Pacific Northwest National Laboratory

www.ingramcontent.com/pod-product-compliance
Lightning Source LLC
Chambersburg PA
CBHW080650180526
45168CB00008B/3372